Oxford Lecture Series in
Mathematics and its Applications 8

Series editors

John Ball   Dominic Welsh

OXFORD LECTURE SERIES IN
MATHEMATICS AND ITS APPLICATIONS

1. J. C. Baez (ed.): *Knots and quantum gravity*
2. I. Fonseca and W. Gangbo: *Degree theory in analysis and applications*
3. P. L. Lions: *Mathematical topics in fluid mechanics, Vol. 1: Incompressible models*
4. J. E. Beasley (ed.): *Advances in linear and integer programming*
5. L. W. Beineke and R. J. Wilson (eds): *Graph connections: Relationships between graph theory and other areas of mathematics*
6. I. Anderson: *Combinatorial designs and tournaments*
7. G. David and S. W. Semmes: *Fractured fractals and broken dreams*
8. Oliver Pretzel: *Codes and algebraic curves*
9. M. Karpinski and W. Rytter: *Fast parallel algorithms for graph matching problems*
10. P. L. Lions: *Mathematical topics in fluid mechanics, Vol. 2: Compressible models*

Codes and Algebraic Curves

Oliver Pretzel

Mathematics Department
Imperial College, London

CLARENDON PRESS • OXFORD
1998

Oxford University Press, Great Clarendon Street, Oxford OX2 6DP

Oxford New York
Athens Auckland Bangkok Bogota Bombay
Buenos Aires Calcutta Cape Town Dar es Salaam
Delhi Florence Hong Kong Istanbul Karachi
Kuala Lumpur Madras Madrid Melbourne
Mexico City Nairobi Paris Singapore
Taipei Tokyo Toronto Warsaw

and associated companies in
Berlin Ibadan

Oxford is a trade mark of Oxford University Press

Published in the United States
by Oxford University Press, Inc., New York

A catalogue record for this book is available from the British Library

Library of Congress Cataloging in Publication Data
(Data available)

ISBN 0 19 850039 4

Typeset by the author using LaTeX

Printed in Great Britain by
Bookcraft Ltd., Midsomer Norton, Avon

For Ivanova and Joshua

PREFACE

The purpose of this book is to provide an introduction to geometric Goppa codes, a class of error-correcting codes with sensationally good parameters. The author of such a text faces a quandary. How should algebraic geometry, a notoriously difficult but beautiful subject, be presented? It is possible to bypass most or all of the subject, but this has two major drawbacks. One can deal only with a limited class of Goppa codes and the techniques applied to them will be rather *ad hoc*. That is the approach of Feng *et al.* (1994). Perhaps their ideas can be expanded into a full-blown treatment of Goppa codes. That would be welcomed by coding theorists who would no longer be obliged to learn algebraic geometry, and also by algebraic geometers, because it would provide new tools for their subject.

If one decides to include algebraic geometry one still has two alternatives. One can regard curves as geometric objects and describe the theory from this point of view. That is done in the books of Fulton (1989) and van Lint and van der Geer (1988). A disadvantage of this method is that there is a lot of topology to discuss in addition to the algebra. Alternatively one can present the algebra that comes from algebraic geometry without discussing curves themselves. That is the approach of Chevalley (1951) and the school of Roquette (Deuring 1973, Stichtenoth 1991). Without doubt that method provides the most elegant and powerful theory, but it is highly abstract and difficult to motivate.

So I have decided on a dual approach. First I shall present a geometric account of algebraic curves with examples but without proving the harder theorems, leading up to an account of Goppa codes and their error processing algorithms. This description is resolutely affine in spirit and thus avoids most of the topological complications of other geometrical presentations. Then enough of the algebraic theory is presented to prove the results that were only stated initially.

Part I is a revision of the last part of my book (Pretzel 1992), which was omitted from the student edition (Pretzel 1996). It begins at an elementary level and gradually becomes more technical. The deep theorems on which Goppa codes rely are presented without proof but with as strong a motivation as I can give. They are then used to construct and analyse the codes, concluding with a full analysis of two error processing algorithms. The first is a simple algorithm, which does not exploit the full designed correction capability of the codes. The second more complicated one, which does correct up to the full capability of the codes, is based on a remarkable insight of Feng and Rao (1993).

I believe that such presentation is extremely important and in some ways more useful to the beginning student than a sequence of formal proofs. At my own first encounter with the Riemann–Roch theorem, in a very formal algebraic geometry course, I had the impression that it was an almost trivial result obtained by juggling appropriate definitions. It was only as I gradually learned its consequences that I came to appreciate its significance.

For this part the reader is assumed to be familiar with the basic concepts of coding theory including the theory of finite fields, as presented in my book (1996), but no further knowledge of algebra is presumed. If you only want to understand Goppa codes and their error processors and are willing to accept the geometric theorems on faith, you will need only this part, though you might also be interested in the very last chapter of the book, which can be read without ploughing through the intervening theoretical chapters.

Part II provides the proofs missing in Part I. It contains the elements of the theory of fields of functions of one variable in the tradition of Chevalley and the school of Roquette omitting the more difficult areas and simplifying a number of the proofs and definitions. It goes far enough to prove the Riemann–Roch theorem and the Plücker formula for the genus.

This part requires some additional material from a standard algebra course. In particular the reader is assumed to be familiar with rings and ideals and one proof requires Zorn's lemma (which is stated, but not proved). More importantly it requires a higher level of 'mathematical maturity', as it is more abstract and the arguments are longer and more intricate. It concludes with a discussion of the construction of good Goppa codes.

If you want to understand the algebra behind the geometry and the codes, then this is the part you will be most interested in. Apart from the last two chapters, it can be read without reading Part I first, but I think Part I will provide motivation making it much easier to read this part. If you wish to learn more, you can then progress to the books cited above.

I would like to thank the readers of my earlier books who wrote to me with comments and corrections. Writing is a solitary task and it is a real pleasure to hear from readers. I am also grateful to my students at Imperial College. Without their reactions to my courses the book would have been much harder to write. Finally I thank my editor at Oxford University Press, Elizabeth Johnston, for her patience and encouragement, Julia Tompson for her support, and the anonymous copy editor for reading my manuscript so carefully.

Imperial College, London O.P.
April 1997 o.pretzel@ic.ac.uk

CONTENTS

TABLES

Tables 2.1 and 2.2 are also reproduced at the back of the book.

Part I

Codes from algebraic curves

1

INTRODUCTION: CURVES AND CODES

The Aegean island of Delos was one of the great centres of worship of Apollo in ancient Greece. His cave temple stands there to this day. According to an apocryphal story an oracle required the Delians to double the size of a cubical altar. They initially responded by placing a second cube next to the first, but this was rejected. A second attempt created a monstrous altar with double the side length of the original. This too was rejected and it became clear that the requirement was to construct a cubic altar of exactly twice the volume of the original. This became known as the Delic problem. Other versions of this story place it in Crete and state that it was King Minos who wished to double the size of the cubic mausoleum of Glaucus. In any case the doubling of the cube became one of the three classic problems of Greek mathematics.

The Greeks regarded a solution of this problem as perfect only if it was achieved using unmarked ruler and compasses alone. No such solution was ever found and later Greek mathematicians almost certainly realized that the problem was not solvable in this sense, although a proof of that was not found until the nineteenth century.

1.1 The Greeks

Not being able to find a perfect solution did not prevent Greek mathematicians from looking for other techniques. In 340 BC Menaechmus while working on the problem discovered the conic sections: the ellipse, hyperbola and parabola. He gave two solutions to the problem, one by finding the point of intersection of a pair of parabolas, the other by finding the intersection parabola and a hyperbola.

Later Diocles solved the problem using a curve that is now known as the cissoid. This curve is a cubic that has the property that one can construct the intersections with a given line through the origin using ruler and compasses. The lines that give segments of length $\sqrt[3]{2}$ cannot themselves be constructed using ruler and compasses, but arbitrary close approximations can.

1.2 The introduction of coordinates

Starting from this problem and other similar ones the Greeks developed a rich collection of curves, but the systematic study of algebraic curves really starts with the introduction of coordinate geometry by René Descartes in the early seventeenth century. He introduced the familiar x- and y-axes and defined lines and curves by equations. Thus the cissoid has the equation $y^2(2-x) = x^3$.

In the eighteenth and early nineteenth centuries the subject underwent major developments. Among the important results were Bézout's formula for the number of points of intersection of two curves and Plücker's formulae for the numbers of multiple points on a curve

1.3 The use of algebraic structures

In the mid nineteenth century the subject underwent a major change associated with the name of Bernhard Riemann. In the 1850s he introduced abstract structures associated with a curve and derived many new results, the most famous of which is the theorem that bears his name and that of his pupil Roch. His work includes the study of Riemann surfaces, which are curves over the complex numbers, and the introduction of the genus. Older results can be translated into this theory. For instance, Bézout's theorem becomes the degree theorem in this setting, while Plücker's results convert into formulae for the genus. Riemann used analytical methods and assumed an unproven result that he called the Dirichlet principle. To avoid this weakness Clebsch started the process of finding algebraic proofs of Riemann's results in the 1860s. His work was continued by M. Noether and notably in this century by André Weil. In this form the theory applies not just to the complex numbers, but to any fields including finite ones. That is the point at which it links with a second strand of ideas.

1.4 The origins of codes

Some 2300 years after the Greeks had started puzzling over their problem of doubling the cube an American mathematician, Richard Hamming, working at Bell Telephone Laboratories in the late 1940s, was troubled by a very different problem. At that time, in the infancy of modern computing machinery, private work had to be submitted for unsupervised batch processing over the weekends. Any error caused the job to be aborted and the next job started. The unfortunate author might have to wait several weeks before the job could be resubmitted. Hamming asked himself how redundant information could be incorporated in the job submission that would allow the automatic correction of reading errors and in response invented the first error-correcting code. Hamming's idea was to extend each 'block' of input information by additional information so that if a relatively small error occurred, not only would its presence be detected, but its location could be determined so that it could be corrected. The enlarged blocks of information corresponding to all possible inputs form the code.

1.5 Codes found by evaluating functions

The importance of Hamming's idea was quickly recognized and many new error-correcting schemes were developed. Perhaps the most successful were the codes invented by Reed and Solomon which used as their code blocks the values of all polynomials up to a fixed degree. Because computers work in binary the polynomials were taken over a finite field and evaluated at all the points of the field taken in some order. It was later recognized that these codes could be described as members of a wider family of codes for which efficient error processing algorithms were available and this led to their widespread adoption.

1.6 Codes and algebraic curves

Reed–Solomon codes have only one drawback. There are not sufficiently many of them. If you decide to use a field of order 256, then a Reed–Solomon code cannot have block size greater than 256. One consequence of Shannon's general theory of information, developed at Bell Telephone Laboratories at the same time as Hamming's codes, is that the best codes have long block lengths. Around 1980 the Russian mathematician N. V. Goppa suggested that instead of evaluating polynomials at points one should evaluate algebraic functions on curves defined over finite fields. He showed how to replace the condition on the degree of the polynomial by algebraic conditions on the functions. Reed–Solomon codes are a special case of his codes when the curve is a straight line. Curves over finite fields can have many more points than straight lines, so that Goppa's codes can be much longer than Reed–Solomon codes. It was soon shown that these codes were theoretically far superior to almost all previously known codes. Very good geometric Goppa codes are now available and so are error processing algorithms, albeit not as efficient as those for Reed–Solomon codes.

1.7 Chapter layout

The aim of the first part of this book is to describe the geometry of algebraic curves and then to use it to develop Goppa's codes. Chapter 2 discusses the definition of a curve and the points it contains. It introduces three example curves that will be used in the following chapters. Then Chapter 3 introduces the functions on curves that correspond to polynomials. The most theoretical chapter, Chapter 4, describes the 'Riemannian' theory of curves. The main theorems are stated and motivated by examples but not proved, as the necessary mathematical tools have not been developed. Those tools are not needed for the construction and error processing of Goppa codes.

This part concludes with three chapters on Goppa codes. Chapter 5 introduces the codes with several examples and gives estimates for their parameters. Chapter 6 introduces a simple basic error processor. It is straightforward to implement and to understand, but does not exploit the full strength of the codes. The final chapter presents a full error processor based on the majority voting principle introduced by Feng and Rao (1993).

The second part of the book contains an introduction to the theory of function fields. It develops the theory far enough to prove the theorems used in Part I. It gives an axiomatic treatment and can be read independently, but it draws on Part I for motivation and examples. An outline of its content is given in Chapter 8.

1.8 Notational conventions

Each chapter consists of a short introduction followed by a sequence of numbered paragraphs. Thus Chapter 4 begins with Paragraph 4.1, and continues with Paragraph 4.2 and so on. Definitions, theorems, and propositions are unnumbered and referred to by the number of the paragraph in which they lie (a paragraph never contains more than one theorem or proposition). The only exception to this rule is the numbering of exercises. These form the last paragraph of each chapter and are numbered individually.

I use a sans serif fount for symbols that have a constant meaning throughout the text. Thus Z, Q, R, and C are the integers, rational numbers, real numbers, and complex numbers respectively, and F_q is the finite field of order q. The sans serif letter i denotes the number $\sqrt{-1}$ in C and sans serif e the base of the natural logarithms.

Some sections of the text are indented and printed in a smaller typeface. These sections are in some sense optional and can usually be omitted at first reading. Mostly, the sections deal with material that goes further than is required for the main purposes of the book, but there are three other reasons why text may be indented in this way. First, the material covered may concern topics (such as Reed–Solomon codes) which are related to the main topics of the text but are not required for its understanding. Secondly, the material may require methods from linear algebra (such as the use of the characteristic polynomial of a matrix) which, though standard, are not covered by this book and are not part of the general prerequisites. Thirdly, the material may be of a technical nature (such as the proof of unique factorization for polynomials in several indeterminates), where the results but not the ideas are important for the rest of the book.

2

ALGEBRAIC CURVES

How should we define an algebraic curve in the plane? Nowadays, it seems natural to define it as the set of points satisfying some algebraic equation.

2.1 Defining a curve

Let K be a field (which you may initially imagine to be the set of real numbers). We denote the set of polynomials in two indeterminates over a field K by $K[x, y]$ and make a tentative definition of a curve as follows.

INITIAL DEFINITION Let $f(x, y)$ be a polynomial in $K[x, y]$, and let C be the set of points (a, b) for which $f(a, b) = 0$. We say C is the *curve* $f(x, y) = 0$ and write C: $f(x, y) = 0$.

This definition needs some polishing. So let us consider some examples.

EXAMPLE Over the real numbers R, the polynomial $x^2 + y^2 - 1$ corresponds to the circle of points at distance 1 from the origin. Similarly, $y - x^2$ corresponds to a parabola symmetric about the y-axis, and the polynomial y corresponds to the x-axis. These are all bona fide curves.

On the other hand, the product $(x^2 + y^2 - 1)(y - x^2)$ defines a circle together with a parabola, and what about the polynomials $x^2 + y^2 + 1$, or $x^2 + 1$? They have no real points at all.

2.2 Irreducible polynomials in $K[x, y]$

To avoid the case when we get two curves, we must ensure that the defining polynomial $f(x, y)$ does not factor. Polynomials in two indeterminates share most of the properties of ordinary polynomials $f(x)$, but there is an important exception. With respect to addition and multiplication, polynomials in two indeterminates behave as well as one could expect. They form an integral domain. We can also define the degree of a polynomial $f(x, y)$ by taking the degree of $x^m y^n$ to be $m + n$. As with ordinary polynomials, the degree of a product of two polynomials $f(x, y)$ and $g(x, y)$ as the sum of their degrees and the degree of their sum is at most equal to the larger of their individual degrees,

$$\deg(f(x, y)g(x, y)) = \deg(f(x, y)) + \deg(g(x, y)),$$
$$\deg(f(x, y) + g(x, y)) \leq \max\{\deg(f(x, y)), \deg(g(x, y))\}.$$

We can also make the usual definition of irreducibility.

DEFINITION A non-constant polynomial $f(x,y)$ in $K[x,y]$ is called *reducible* if it has a factorization

$$f(x,y) = g(x,y)h(x,y) \qquad (*)$$

with $\deg g(x,y) < \deg f(x,y)$ and $\deg h(x,y) < \deg f(x,y)$. If $f(x,y)$ is not reducible, it is called *irreducible.* Thus in any factorization $(*)$ of an irreducible polynomial $f(x,y)$, one of the factors has the same degree as f and the other is a constant. Constants are neither reducible nor irreducible.

EXAMPLE The polynomials x, y and $x - y$ are all irreducible. So is any irreducible polynomial in a single indeterminate x or y alone. There are many more. For instance, the polynomial $x^3y + y^3 + x$ is irreducible over $\mathsf{Z}/2$.

We can still prove that polynomials factor uniquely as products of irreducibles but the proof is harder because it is not always possible to divide one polynomial by another in such a way that the remainder has degree smaller than the divisor. As Euclid's algorithm depends on such division it is no longer available.

EXAMPLE It is not possible to write x as $q(x,y)(x-y)+r$ where $\deg(r) \leq 0$.

To get round that difficulty we need a lemma, due to Gauss. Recall that a *rational function* is the quotient of two polynomials. We consider polynomials in $K[x,y]$ as polynomials in y with coefficients in $K[x]$. They can then be regarded as a subclass of the set of polynomials whose coefficients are rational functions in x. These 'extended' polynomials are polynomials in a single indeterminate over a field. The lemma states that if the coefficients of a product of 'extended' polynomials lie in $K[x]$, then it is possible to clear all the denominators of the coefficients of both factors.

LEMMA (Gauss) *Let K be a field. Consider the polynomial ring $K[x]$ and let $R = K(x)$ be the field of rational functions in the indeterminate x. Let $p(y)$ and $q(y)$ be polynomials in $R[y]$ and suppose that $p(y)q(y)$ has all its coefficients in $K[x]$; then there exists a non-zero element $r(x) \in R$ such that $r(x)p(y)$ and $r^{-1}(x)q(y)$ have all their coefficients in $K[x]$.*

Proof The proof uses unique factorization in $K[x]$ as its basic tool.
(1) We first bring the coefficients of $p(x)$ and $q(x)$ over a common denominator so that we have

$$p(y) = \sum \frac{a_i(x)y^i}{c(x)} \qquad q(y) = \sum \frac{b_j(x)y^j}{d(x)}.$$

We assume that we have already multiplied $p(x)$ by $r(x)$ and $q(x)$ by $r^{-1}(x)$ and cancelled common factors so that the degree of $c(x)d(x)$ is as small as possible. If that degree is 0, then both $c(x)$ and $d(x)$ are constants in K and so all the coefficients of $p(y)$ and $q(y)$ lie in $K[x]$.

(2) Suppose therefore that the degree of one of $c(x)$ and $d(x)$ is at least 1, say $\deg c(x) > 0$. Let $t(x)$ be an irreducible factor of $c(x)$. Then by hypothesis there must be some i for which $t(x)$ does not divide $a_i(x)$. For otherwise, we could cancel the common factor $t(x)$ from all coefficients $a_i(x)$ and $c(x)$, reducing its degree.

If, furthermore, $t(x)$ divides every coefficient $b_j(x)$, then we can multiply $p(y)$ by $t(x)$ and divide $q(y)$ by $t(x)$. That reduces the degree of $c(x)$ but does not change $d(x)$, since $t(x)$ cancels with all the coefficients $b_j(x)$. As we chose the degree of $c(x)d(x)$ to be minimal, this cannot happen. Hence there must be some j for which $t(x)$ does not divide $b_j(x)$.

(3) Now we derive a contradiction. Let i and j be the smallest indices for which $t(x)$ does not divide $a_i(x)$ and $b_j(x)$ respectively and put $k = i + j$. The coefficient of y^k in $p(y)q(y)$ is

$$\frac{\sum_{l=0}^{k} a_l(x)b_{k-l}(x)}{c(x)d(x)}.$$

If this lies in $K[x]$, then $t(x)$ must divide the numerator, but that can be split up in the following way:

$$\sum_{l=0}^{k} a_l(x)b_{k-l}(x) = a_i(x)b_j(x) + \sum_{l<i} a_l(x)b_{k-l}(x) + \sum_{l<j} a_{k-l}(x)b_l(x).$$

On the right hand side $t(x)$ divides each term $a_l(x)$ in the first sum and each term $b_l(x)$ in the second. So both sums are divisible by $t(x)$. On the other hand $t(x)$ is irreducible and does not divide either of $a_i(x)$ or $b_j(x)$. Thus it does not divide their product. Hence $t(x)$ does not divide the right hand side of the equation, contradicting the assumption that it divides the sum on the left hand side.

This contradiction shows that the hypothesis that the degree of $c(x)d(x)$ is greater than 0 is untenable. Therefore $c(x)$ and $d(x)$ are constants and the lemma is proved. □

2.3 Unique factorization in $K[x,y]$

We are now able to prove that every polynomial in $K[x,y]$ factors uniquely into irreducibles. This paragraph contains a proof of the key point, namely that if an irreducible divides a product, then it also divides one of the factors. The rest of the proof follows exactly the lines of the proof for polynomials in a single indeterminate and is left to the exercises (Exercise 1).

PROPOSITION *If $f(x,y)$ is an irreducible polynomial and $f(x,y)$ divides the product $g(x,y)h(x,y)$, then $f(x,y)$ divides one of $g(x,y)$ and $h(x,y)$.*

Proof (1) Most irreducible polynomials of $K[x,y]$ do not lie in $K[x]$ and certainly none lie in both $K[x]$ and $K[y]$ because the only common elements are the elements of K and they are not irreducible. We assume therefore that $f(x,y)$ does not lie in $K[x]$ and introduce the field R of rational functions in x.

By collecting terms with like powers of y, f, g, and h can be considered as polynomials in $R[y]$ which obeys the usual rules for polynomials over a field.

(2) The next step is to show that f is still irreducible in $R[y]$. Suppose that $f = p(y)q(y)$ in $R[y]$. Then by Gauss' lemma we can multiply $p(y)$ by $r(x)$ and $q(y)$ by $r^{-1}(x)$ so that they lie in $K[x,y]$. But then one of the resulting polynomials is a constant because $f(x,y)$ is irreducible in $K[x,y]$. Hence one of $p(y)$ or $q(y)$ lies in R, proving that $f(x,y)$ is irreducible in $R[y]$.

(3) Now we use the ordinary theory of polynomials in a single indeterminate to establish that f divides one of g and h in $R[y]$. Thus, say $g(x,y) = f(x,y)q(y)$ for some $q(y) \in R[y]$.

Applying Gauss' lemma to this product, it follows that for some rational function $s(x) = a(x)/b(x)$ in cancelled form, both $f(x,y)s(x)$ and $s^{-1}(x)q(y)$ lie in $K[x,y]$. The polynomials $b(x)$ and $a(x)$ are relatively prime. So $b(x)$ must divide every coefficient of f considered as a polynomial over $K[x]$. Hence it must divide $f(x,y)$ which is irreducible. Therefore $b(x)$ is a constant or a constant multiple of $f(x,y)$.

(4) It is impossible for $b(x)$ to be a constant multiple of $f(x,y)$ because, by assumption, $f(x,y)$ does not lie in $K[x]$. So $b(x)$ is a constant b. Then $s^{-1}q(y) = bq(y)/a(x)$ lies in $K[x,y]$ and hence so does $q(y) = (a(x)/b)(bq(y)/a(x))$. Hence $f(x,y)$ divides $g(x,y)$ in $K[x,y]$. □

It now follows by the standard arguments that every polynomial in two indeterminates has a unique factorization into irreducibles. With the aid of unique factorization we can still find highest common factors (HCFs), but it is no longer true that the HCF of $f(x,y)$ and $g(x,y)$ can be written in the form $uf + vg$.

EXAMPLE The HCF of x and y is 1.

Any polynomial that divides x must have degree at most 1. So it must have the form $ax+by+c$. If the product of two such polynomials is x then comparing coefficients one sees that one must have the form ax and the other must be the constant $1/a$. This shows that x is, as to be expected, irreducible. As x does not divide y it follows that the HCF of x and y is 1.

On the other hand any polynomial of the form $xs(x,y) + yt(x,y)$ has zero constant term, so it cannot be 1.

From this it follows that Euclid's algorithm and hence division with remainder cannot be valid for *any* degree function on polynomials in two indeterminates. It is possible to find a partial substitute in the theory of Gröbner bases (as described in Cox *et al.* 1992). Gröbner basis theory underlies most algorithms for computer calculations involving polynomials in several indeterminates. We shall only make theoretical use of such calculations and so do not need to discuss that theory here.

2.4 Increasing the supply of curves

From now on, when we speak of a curve $C : f(x,y) = 0$ we shall assume that the defining polynomial $f(x,y)$ is irreducible. That eliminates the possibility that C splits into two curves.

A plane curve over F_2 cannot have more than four points with coefficients in F_2 itself, because the only available points are (0,0), (0,1), (1,0), (1,1). This means that, as far as the points over F_2 alone are concerned, there are only at most $2^4 = 16$ plane curves. That is rather restrictive. To get enough different curves, we must admit points with coordinates (a,b) in fields L containing K. For instance, if we define curves over the real numbers, we shall still regard the complex point $(\mathrm{i}, \sqrt{2})$ as part of the unit circle (because $\mathrm{i}^2 + 2 = 1$). We do not need to allow all possible fields containing K, and restrict our attention to finite extensions.

DEFINITION A field L is called a *finite extension* of K if L is finite dimensional as a vector space over K.

EXAMPLE The complex numbers C form a finite extension of the real numbers R (with basis 1, i). If $K \subseteq L$ are finite fields, then L is a finite extension of K. The real numbers R do not form a finite extension of the rational numbers Q.

In order to have sufficiently many curves, we admit points with coordinates in finite extension fields.

2.5 Absolute irreducibility

When we extend from K to L it is possible that the polynomial defining our curve C: $f(x, y) = 0$ may factor.

EXAMPLE Suppose we take $K = \mathsf{R}$ and consider the 'curve' $C : x^2 + 1 = 0$. This has no real points at all, but over C it splits into two straight lines, $x = \mathsf{i}$ and $x = -\mathsf{i}$.

On the other hand the curve $C : x^2 + y^2 + 1 = 0$, which also has no real points, becomes a 'circle of radius i' with points such as $\mathsf{i}(a, b)$, where (a, b) lies on the circle $x^2 + y^2 - 1 = 0$.

To avoid the problem with x^2+1, we now require that the polynomial defining our curve remains irreducible over every extension field we consider. To be completely safe we can require it to be absolutely irreducible in the following sense.

DEFINITION A non-constant polynomial $f(x, y)$ in $K[x, y]$ is called *absolutely irreducible* if $f(x, y)$ is irreducible in $L[x, y]$ for every finite extension L of K.

The only polynomials of the form $f(x)$ that are absolutely irreducible have the form $f(x) = ax + b$ because every polynomial $f(x)$ splits into linear factors in some finite extension field.

The following proposition and its corollary show that for polynomials in two indeterminates absolutely irreducible polynomials behave rather as linear polynomials do for a single indeterminate.

PROPOSITION *Let $f(x, y)$ be a non-constant polynomial over a field K. Then there exists a finite extension field L such that $f(x, y)$ has an absolutely irreducible factor in $L[x, y]$.*

Proof The proof is by induction on the degree of $f(x, y)$. If that is 1, then $f(x, y)$ is absolutely irreducible itself. Suppose that the proposition has been proved for all degrees less than n and that $f(x, y)$ has degree n.

If $f(x, y)$ is absolutely irreducible than there is nothing to show. Otherwise there exists a finite extension field J such that $f(x, y)$ is not irreducible in $J[x, y]$ (of course, J may be K). Then in $J[x, y]$,

$$f(x, y) = g(x, y)h(x, y)$$

where both g and h have degree less than n and neither is a constant. By our induction hypothesis it follows that for some finite extension field L of J, $g(x,y)$ has an absolutely irreducible factor $p(x,y)$ in $L[x,y]$. Now $p(x,y)$ is also a factor of $f(x,y)$ in $L[x,y]$ and L is a finite extension of K. So the proposition holds for $f(x,y)$ of degree n. □

COROLLARY *Let $f(x,y)$ be a non-constant polynomial over a field K. Then there exists a finite extension field L such that $f(x,y)$ splits into a product of absolutely irreducible factors in $L[x,y]$.*

Proof The proof is again by induction on the degree n of $f(x,y)$. If $f(x,y)$ is itself absolutely irreducible there is nothing to prove. That is always the case if $n = 1$.

If, on the other hand, $f(x,y)$ is not absolutely irreducible, then by the proposition there exists a finite extension field J of K such that

$$f(x,y) = p(x,y)q(x,y) \qquad (*)$$

in $J[x,y]$ and $p(x,y)$ is absolutely irreducible. Then $\deg q(x,y) < n$ and so by the induction hypothesis there exists a finite extension field L of J such that $q(x,y)$ splits into absolutely irreducible factors over L. Then the same holds for $f(x,y)$ because of eqn $(*)$. □

2.6 Existence of absolutely irreducible polynomials

The following proposition, which is a special case of Eisenstein's criterion, gives an easy way to construct absolutely irreducible polynomials. The criterion is by no means necessary, but there are sufficiently many polynomials that satisfy it (including all our examples) for it to be useful.

PROPOSITION Eisenstein's criterion *Let $f(x,y) \in K[x,y]$ be written in the form*

$$f(x,y) = \sum_{i=0}^{n} f_i(x)y^i.$$

Suppose the highest common factor of $f_0(x), \ldots, f_n(x)$ is 1 and that there exists an element $a \in L$ for some finite extension field of K such that

(i) *a is not a root of $f_n(x)$;*

(ii) *a is a root of $f_i(x)$ for all $i < n$; and*

(iii) *a is not a double root of f_0.*

Then $f(x,y)$ is absolutely irreducible.

Proof Suppose that for some finite extension J of K, the polynomial $f(x,y)$ factors non-trivially. Let $f(x,y) = g(x,y)h(x,y)$, where $g(x,y) = \sum^r g_i(x)y^i$ and $h(x,y) = \sum^s h_i(x)y^i$, and $0 < r, s \le n$. Then that factorization will also be non-trivial in any finite extension of J. So we may assume that $a \in J$.

As a is not a double root of $f_0(x) = g_0(x)h_0(x)$ it is a root of one of $g_0(x)$ and $h_0(x)$ but not the other. Say $g_0(a) = 0 \neq h_0(a)$. Further, since a is not a root of $f_n(x) = g_r(x)h_s(x)$ it is not a root of either $g_r(x)$ or $h_s(x)$. Now let

$k > 0$ be the smallest index such that a is not a root of $g_k(x)$ and consider the coefficient

$$f_k(x) = g_k(x)h_0(x) + \sum_{i=0}^{k-1} g_i(x)h_{k-i}(x).$$

If $k < n$, a is a root of $f_k(x)$. It is also, by the definition of k, a root of $g_i(x)$ for every $i < k$, but it is not a root of $g_k(x)$ or $h_0(x)$. So

$$0 = f_k(a) = g_k(a)h_0(a) + \sum_{i=0}^{k-1} g_i(a)h_{k-i}(a) = g_k(a)h_0(a) + 0 \neq 0.$$

This contradiction proves that $k = n$. As $f(x, y)$ has no coefficients $f_i(x)$ with $i > n$, it follows that $h(x, y) = h_0(x)$. Hence $h_0(x)$ divides every polynomial $f_i(x)$ for $i = 1, \ldots, n$. The fact that these have highest common factor 1 implies that $h_0(x)$ is a constant. Thus in any factorization of $f(x, y)$, over any finite extension of J, one of the factors is a constant, showing that $f(x, y)$ is absolutely irreducible. □

COROLLARY *There are infinitely many, distinct, absolutely irreducible polynomials over any field.*

Proof By Eisenstein's criterion (with $a = 0$) all polynomials of the form $y^n - x$ are absolutely irreducible □

2.7 Projective transformations

Allowing curves to have points in finite extensions is still not quite sufficient. It is necessary to consider another kind of point. You will probably be familiar with the fact that for many theorems about functions on complex numbers one must admit an additional 'point at infinity'. That point ensures that all rational functions including $1/z$ have zeros. For our theory we need to allow points at infinity for exactly the same reason. We can obtain these by allowing certain non-linear coordinate transformations which are called 'projective' because of their relation to perspective drawing. We shall need just two such transformations:

DEFINITION The coordinate transformations from the standard (x, y)-system to the (u, v)- and (w, z)-systems given by the rules $u = y/x$, $v = 1/x$ and $w = 1/y$, $z = x/y$ will be called projective coordinate changes.

The transformation from the (u, v)-system to the (w, z)-system is of the same type (see Exercise 4). Points can be given in any of the three coordinate systems. Most are representable in all three systems, but some are only representable in two of them and a very few are only representable in one system.

EXAMPLE The point $(x{=}1, y{=}1)$ is the same as the point $(u{=}1, v{=}1)$, and the point $(x{=}a, y{=}b)$ for $a \neq 0$ is the same as the point $(u{=}ba^{-1}, v{=}a^{-1})$. The point $(x{=}0, y{=}1)$ has no equivalent in the (u, v)-system, but it is the same as $(w{=}1, z{=}0)$. The point $(u{=}1, v{=}0)$ has no equivalent in the (x, y)-system, but it is the same as $(w{=}0, z{=}1)$. The point $(x{=}0, y{=}0)$ has no equivalent in the (u, v)- or the (w, z)-system.

CONVENTION We usually specify the coordinate system we are using by the notation P: $(x{=}a, y{=}b)$. The systems are used in the following order of priority. If a point can be represented in the (x, y)-system, then we use that system. Failing that, if the point can be represented in the (u, v)-system (necessarily with $v = 0$) we use that system. The single remaining point $(w{=}0, z{=}0)$ is represented in the (w, z)-system.

We can consider the points that cannot be represented in the (x, y)-system as forming a *line at infinity* or *horizon* with the equation $v = 0$. The horizon intersects every ordinary line in a single point. Thus, as we shall shortly verify, $(u{=}a, v{=}0)$ is the intersection of the line $y = ax + b$ with the horizon, and $(w{=}0, z{=}0)$ is the intersection of the y-axis $(x = 0)$ with the horizon.

2.8 Final definition of a curve

To decide whether a point $(u{=}b, v{=}0)$ lies on a curve C: $f(x, y) = 0$, we need to transform the equation itself. That can be done provided $f(x, y) \neq ax$. In the exceptional case the curve is a straight line. That line is, of course, the horizon of the (u, v)-system.

EXAMPLE The equation of the straight line $y = ax + b$ can be written as $y/x = a + b/x$. In the (u, v)-system that transforms to $u = a + bv$, confirming that $(u{=}a, v{=}0)$ lies on the line.

The rectangular hyperbola $x^2 - y^2 = 1$ transforms to $1/v^2 - u^2/v^2 = 1$. For $x \neq 0$ also $1/v \neq 0$. So we can multiply this equation by v^2 to get the curve $1 = u^2 + v^2$. This is the equation of a circle with the line $v = 0$ as a diameter. So the rectangular hyperbola can be viewed as a circle with the horizon as a diameter.

The parabola $2x = y^2 + 1$ transforms to $2/v = u^2/v^2 + 1$. Again we multiply by v^2 to get $2v = u^2 + v^2$ or $u^2 + (v - 1)^2 = 1$. This is also the equation of a circle, but this time $v = 0$ is a tangent. So the parabola can be viewed as a circle with the horizon as a tangent.

The curve $x^3 + y^3 - 1 = 0$ transforms to $u/v^3 + 1^3/v^3 - 1 = 0$ or $u^3 + 1 - v^3 = 0$ the same way. If, as will be the case in our examples, the characteristic of the underlying field is 2, then the formula remains the same as the original formula in x and y.

Notice that in all cases the degree of the equation is not changed by the transformation.

We can now give our final definition of an algebraic curve.

DEFINITION Let $f(x, y)$ be an absolutely irreducible polynomial in $K[x, y]$ of degree d and define

$$g(u, v) = v^d f(1/v, u/v);$$
$$h(w, z) = w^d f(z/w, 1/w).$$

Then the *projective* curve C defined by $f(x,y)$ is the union of the three curves C_1: $f(x,y)=0$, C_2: $g(u,v)=0$, and C_3: $h(w,z)=0$, points being allowed to have coordinates in any finite extension L of K. We call these three curves the *affine components* of C, and will still denote the full curve by C: $f(x,y)=0$.

To restrict attention to a particular affine component we speak of the *affine curve* C: $f(x,y)=0$.

If $f(x,y)=ax$, then $g(u,v)=a$, so the (u,v)-affine component is empty. That is because $f(x,y)$ is then the horizon of the (u,v)-system. For other cases we must show that the affine components fit together properly, and that g and h satisfy the conditions for an affine curve.

PROPOSITION *Let $f(x,y) \neq ax$ be an absolutely irreducible polynomial of degree d and let $g(u,v)=v^d f(1/v,u/v)$. Then g is an absolutely irreducible polynomial of degree d. Furthermore, if a point P has coordinates P: $(x{=}a, y{=}b)$ and also P: $(u{=}c, v{=}d)$, then $f(a,b)=0$ if and only if $g(c,d)=0$.*

Proof Corresponding to a term $bx^r y^s$ of $f(x,y)$, the polynomial $g(u,v)$ has a term $bu^s v^{d-r-s}$. As $f(x,y) \neq ax$ and $f(x,y)$ is absolutely irreducible, x does not divide $f(x,y)$. Hence at least one term of $f(x,y)$ has $r=0$. That proves that g has degree d. Further, since the degree of $f(x,y)$ is d there exists a non-zero term $bx^r y^s$ with $r+s=d$. That term transforms to bu^s. Hence $g(u,v)$ is not a multiple of v.

Suppose that, over some finite extension field, $g(u,v)=h(u,v)k(u,v)$, with h and k of degrees r and s respectively. Then neither $h(u,v)$ nor $k(u,v)$ is a multiple of v. So

$$f(x,y)=x^d g(y/x,1/x)=\left(x^r h(y/x,1/x)\right)\left(x^s k(y/x,1/x)\right),$$

and the two factors are polynomials of degree r and s respectively. Since f is absolutely irreducible, it follows that $r=0$ or $s=0$. But that implies that one of h and k is a constant. So g is absolutely irreducible.

We turn to the statement on P. By assumption $c=b/a$ and $d=1/a$ and $a \neq 0$. Then $g(c,d)=a^{-d}f(a,b)$. Thus $g(c,d)=0$ if and only if $f(a,b)=0$.

□

2.9 Curves over finite fields

Until now our examples have been real curves, to aid the reader's geometrical intuition, but for codes we need to use curves over finite fields. I conclude the chapter with three such curves. These examples will be continued in later chapters. For all three curves we shall calculate all the points with coordinates in the fields F_2, F_4, F_8, and F_{16}. F_2 and F_4 are subfields of F_{16}, but F_8 is not. So we construct tables of the F_{16} and F_8. For a detailed explanation of the construction of finite fields in general and these two fields in particular the reader is referred to my book (1996).

2.9.1 *The field F_{16} based on x^4+x^3+1*

The elements of F_{16} are polynomials of degree at most 3 over $\mathsf{F}_2=\{0,1\}$.

Table 2.1 The field F_{16} based on $x^4 + x^3 + 1$

log	–	0	1	12	2	9	13	7	3	4	10	5	14	11	8	6
	0	**1**	**2**	**3**	**4**	**5**	**6**	**7**	**8**	**9**	**10**	**11**	**12**	**13**	**14**	**15**
0	× 0	0	0	0	0	0	0	0	0	0	0	0	0	0	0	0
1	+ 1	1	2	3	4	5	6	7	8	9	10	11	12	13	14	15
2	2	3	4	6	8	10	12	14	9	11	13	15	1	3	5	7
3	3	2	1	5	12	15	10	9	1	2	7	4	13	14	11	8
4	4	5	6	7	9	13	1	5	11	15	3	7	2	6	10	14
5	5	4	7	6	1	8	7	2	3	6	9	12	14	11	4	1
6	6	7	4	5	2	3	13	11	2	4	14	8	3	5	15	9
7	7	6	5	4	3	2	1	12	10	13	4	3	15	8	1	6
8	8	9	10	11	12	13	14	15	15	7	6	14	4	12	13	5
9	9	8	11	10	13	12	15	14	1	14	12	5	8	1	3	10
10	10	11	8	9	14	15	12	13	2	3	11	1	5	15	8	2
11	11	10	9	8	15	14	13	12	3	2	1	10	9	2	6	13
12	12	13	14	15	8	9	10	11	4	5	6	7	6	10	7	11
13	13	12	15	14	9	8	11	10	5	4	7	6	1	7	9	4
14	14	15	12	13	10	11	8	9	6	7	4	5	2	3	2	12
15	15	14	13	12	11	10	9	8	7	6	5	4	3	2	1	3

The number n represents its binary 4-tuple a, b, c, d, which in turn represents the polynomial $ax^3 + bx^2 + cx + d$. Thus $13 \sim 1, 1, 0, 1 \sim x^3 + x^2 + 1$. Addition (the lower half of Table 2.1) is ordinary addition over F_2. In particular $a + a = 0$ for any a, so this does not appear in the table. Multiplication (the upper half of Table 2.1) is multiplication in $\mathsf{F}_2[x]$, followed if necessary by taking the remainder after division by $x^4 + x^3 + 1$.

Every element is a power of $2 \sim x$, and this provides 'logarithms'. Using the fact that $2^{15} = 1$ we can multiply easily and also find inverses easily. To multiply, take logarithms, add them modulo 15 and find the entry that has the answer as its logarithm. For example, 3 and 9 have logarithms 12 and 4 respectively. Adding modulo 15 gives 1, and the entry with 1 as its logarithm is 2. Thus $3 \cdot 9 = 2$. The inverse of the element with logarithm m is the element with logarithm $15 - m$.

The field F_4 is the subfield of F_{16} consisting of the elements $\{0, 1, 10, 11\}$.

2.9.2 *The field* F_8 *based on* $x^3 + x^2 + 1$

The construction of this field, represented in Table 2.2, is entirely analogous to that of F_{16}, except that the field contains only the polynomials of degree at most 2, and we divide modulo $x^3 + x^2 + 1$. To avoid confusion we use $2'$, $3'$, etc., when referring to the elements of this field in the text. Again the lower half of the table represents addition and the upper half represents multiplication. As $2'^7 = 1$ we add modulo 7 when using logarithms.

Table 2.2 The field F_8 based on $x^3 + x^2 + 1$

log		–	0	1	5	2	3	6	4
		0	**1**	**2**	**3**	**4**	**5**	**6**	**7**
0	×	0	0	0	0	0	0	0	0
1	+	1	1	2	3	4	5	6	7
2		2	3	4	6	5	7	1	3
3		3	2	1	5	1	2	7	4
4		4	5	6	7	7	3	2	6
5		5	4	7	6	1	6	4	1
6		6	7	4	5	2	3	3	5
7		7	6	5	4	3	2	1	2

2.10 An example of a cubic curve

In this and the following examples we shall observe the convention that we use the (x, y)-system where possible, the (u, v)-system for points of the form $(u, v{=}0)$, and the (w, z)-system only for $(w{=}0, z{=}0)$.

EXAMPLE The cubic curve $x^3 + y^3 + 1 = 0$.

The polynomial is absolutely irreducible over F_2 by Eisenstein's criterion with the root $a = 1$. The curve has the equation $u^3 + v^3 + 1 = 0$ in the (u, v)-system and equation $w^3 + z^3 + 1 = 0$ in the (w, z)-system.

The points of the curve are listed in Table 2.3. Notice that all the points over F_{16} already have coordinates in F_4. That is because $x^3 = 0, 1, 3, 5, 8$, or 15. The only two cubes adding to 1 are 0 and 1, and the only solutions of $x^3 = 1$ are 1, 10, and 11.

2.11 The Klein quartic

EXAMPLE The Klein quartic has equation $x^3y + y^3 + x = 0$. The polynomial is absolutely irreducible over F_2 by Eisenstein's criterion using the root $a = 0$. It has $u^3v + v^3 + u = 0$ and $w^3z + z^3 + w = 0$ as its equations in the other coordinate systems.

Table 2.3 Points of the cubic $x^3 + y^3 + 1 = 0$

Field	Points
F_2	$(x{=}0, y{=}1)$; $(x{=}1, y{=}0)$; $(u{=}0, v{=}1)$.
F_4	points over F_2 and $(x{=}0, y{=}10), (x{=}0, y{=}11)$; $(x{=}10, y{=}0), (x{=}11, y{=}0)$; $(u{=}10, v{=}0), (u{=}11, v{=}0)$.
F_8	points of F_2 and $(x{=}2', y{=}5'), (x{=}4', y{=}6'), (x{=}7', y{=}3')$; $(x{=}5', y{=}2'), (x{=}6', y{=}4'), (x{=}3', y{=}7')$.
F_{16}	points over F_4 and no further points.

Table 2.4 Points of the Klein quartic $x^3y + y^3 + x = 0$

Field	Points
F_2	$(x{=}0, y{=}0)$; $(u{=}0, v{=}0)$; $(w{=}0, z{=}0)$.
F_4	points over F_2 and
	$(x{=}10, y{=}11)$, $(x{=}11, y{=}10)$.
F_8	points over F_2 and
	$(x{=}1', y{=}3')$, $(x{=}1', y{=}5')$, $(x{=}1', y{=}6')$;
	$(x{=}3', y{=}1')$, $(x{=}5', y{=}1')$, $(x{=}6', y{=}1')$;
	$(x{=}2', y{=}2')$, $(x{=}4', y{=}4')$, $(x{=}7', y{=}7')$;
	$(x{=}2', y{=}5')$, $(x{=}4', y{=}6')$, $(x{=}7', y{=}3')$;
	$(x{=}2', y{=}7')$, $(x{=}4', y{=}2')$, $(x{=}7', y{=}4')$;
	$(x{=}3', y{=}4')$, $(x{=}5', y{=}7')$, $(x{=}6', y{=}2')$;
	$(x{=}3', y{=}5')$, $(x{=}5', y{=}6')$, $(x{=}6', y{=}3')$.
F_{16}	points over F_4 and
	$(x{=}3, y{=}10)$, $(x{=}5, y{=}11)$, $(x{=}8, y{=}10)$, $(x{=}15, y{=}11)$;
	$(x{=}10, y{=}2)$, $(x{=}11, y{=}4)$, $(x{=}10, y{=}9)$, $(x{=}11, y{=}14)$;
	$(x{=}6, y{=}8)$, $(x{=}13, y{=}15)$, $(x{=}7, y{=}3)$, $(x{=}12, y{=}5)$.

The points are listed in Table 2.4. Notice that once a single point (a, b) on the curve has been found, the point (a^2, b^2) also lies on the curve. The classes of points obtained this way are separated by semi-colons.

2.12 A quintic

EXAMPLE The quintic curve $x^5 + y^5 + 1 = 0$.

The polynomial is absolutely irreducible by Eisenstein's criterion using the root $a = 1$. The equations of the curve in the other systems are $u^5 + v^5 + 1 = 0$ and $w^5 + z^5 + 1 = 0$.

The points are listed in Table 2.5. There are three points over F_2 of the form $(1, 0)$ in an appropriate coordinate system. We obtain two further points over F_4 by noting that $10^5 = 10$ and $11^5 = 11$. The remaining 60 points over F_{16} are obtained as follows.

The fifth power of any element of F_{16} is one of the 0, 1, 10, or 11. From the equation of the curve a point must have coordinates whose fifth powers are 0 and 1 or 10 and 11. There are five fifth roots of 1: 1, 3, 5, 8, and 15. Points of the form (0,1) have already been counted, but there are 12 more: $(x{=}0, y{=}3, 5, 8, 15)$, $(x{=}3, 5, 8, 15, y{=}0)$, $(u{=}3, 5, 8, 15, v{=}0)$. This notation means that in each case any value for the first coordinate can be paired with any value for the second.

There are also five fifth roots of 10: 10, 4, 9, 12, 14; and five fifth roots of 11: 11, 2, 6, 7, 13. The points these produce can all be expressed in the (x, y)-system because they have no zero coordinates. The two sets of 25 points we get are $(x{=}10, 4, 9, 12, 14, y{=}11, 2, 6, 7, 13)$ and $(x{=}1, 2, 6, 7, 13, y{=}10, 4, 9, 12, 14)$.

The two points (10,11) and (11,10) have already been counted. The 48 remaining points fall into classes of four, obtained by the squaring method as in the previous example.

Table 2.5 Points of the quintic $x^5 + y^5 + 1 = 0$

Field	Points
F_2	$(x{=}0, y{=}1)$; $(x{=}1, y{=}0)$; $(u{=}1, v{=}0)$.
F_4	points over F_2 and
	$(x{=}10, y{=}11)$, $(x{=}11, y{=}10)$.
F_8	points over F_2 and
	$(x{=}2', y{=}5')$, $(x{=}4', y{=}6')$, $(x{=}7', y{=}3')$;
	$(x{=}5', y{=}2')$, $(x{=}6', y{=}4')$, $(x{=}7', y{=}3')$.
F_{16}	points over F_4 and 60 further points:
	$(x{=}0, y{=}3, 5, 8, 15)$;
	$(x{=}3, 5, 8, 15, y{=}0)$;
	$(u{=}3, 5, 8, 15, v{=}0)$;
	$(x{=}10, 4, 9, 12, 14, y{=}11, 2, 6, 7, 13)$;
	$(x{=}1, 2, 6, 7, 13, y{=}10, 4, 9, 12, 14)$.

The main purpose of this example is to show that curves can have a large number of points. The existence of curves with many points is the underlying reason why there are such good geometric Goppa codes. This is an example of a Hermitian curve. These curves are defined by polynomials of degree $q+1$ over fields F_{q^2}. Codes derived from such curves are among the best codes obtainable from plane curves. They have been extensively studied and will form the subject of a number of exercises.

2.13 Exercises

1. Using the results proved in this chapter, show that every polynomial $f(x, y)$ in $K[x, y]$ can be factored into irreducibles and that this factorization is unique up to order and multiplication of the terms by non-zero constants.

2. Show that $y^2 - x$ is absolutely irreducible over R.

3. Show that $x^2 + y^2 - 1$ is not irreducible over F_2 and that $y^2 - x$ is absolutely irreducible over F_2 (both from first principles and using Eisenstein's criterion).

4. Show that the transformation from the (u, v)-system to the (w, z)-system is given by $w = v/u$, $z = 1/u$.

5. Homogeneous coordinates. Replace an (x, y)-point in the plane by the point $(x, y, 1)$, a (u, v)-point by $(1, u, v)$, and a (w, z)-point by $(z, 1, w)$. Now regard two points (a, b, c) and (a', b', c') as equivalent if there exists a non-zero constant k with $a' = ka$, $b' = kb$, and $c' = kc$. Show that (a, b, c) is equivalent to an (x, y)-point if and only if $c \neq 0$. Show that the (x, y)-point equivalent to (a, b, c) is unique if it exists. Finally, show that the points in the (x, y)- and (u, v)-points equivalent to (a, b, c) (where $a \neq 0$ and $c \neq 0$) are linked by the projective transformation of the text.

6. Why did I choose the Klein quartic as our example of a curve of degree 4, and not the curve $x^4 + y^4 = 1$?

7. Find the points of the curve $x^5 + y^4 + y = 0$ in the fields F_2, F_4, F_8, and F_{16}.

8. Find the points of the curve $x^7 + y^7 = 1$ in the fields F_2, F_4, F_8, and F_{16}.

9. Hermitian curves. Let K be the field F_{q^2}. Consider the two Hermitian curves C: $x^{q+1} + y^{q+1} + 1$ and $D : x^{q+1} - y^q - y$. Calculate their equations in the (u, v)- and (w, z)-systems and show that they each have $q^3 + 1$ points over K. Show that D has a unique point on the horizon of the (x, y)-system.

The next two exercises will show that these two curves are in some sense the same (they provide a birational equivalence between them).

10. Let p be a prime number and let $q = p^k$. Show that

$$(t^q + t + 1)^q - (t^q + t + 1) = t^{q^2} - t$$

in $\mathsf{F}_p[t]$. Deduce that $K = \mathsf{F}_{q^2}$ contains the roots of $t^q + t + 1$. Show also that K contains the roots of $t^{q+1} + 1$.

11. Hermitian curves (cont.). Let $a, b \in K = \mathsf{F}_{q^2}$ be roots of $t^q + t + 1$ and $t^{q+1} + 1$ respectively. Show that if $x^{q+1} + y^{q+1} + 1 = 0$, then $bx \neq y$.

Let $x' = b/(y - bx)$ and $y' = xx' - a$. Show that

$$(y - bx)(y'^q + y' - x'^{q+1}) = 0$$

and deduce that $x'^{q+1} - y'^q - y' = 0$.

3

FUNCTIONS ON ALGEBRAIC CURVES

In the previous chapter we obtained a satisfactory definition of a plane algebraic curve. The next step is to investigate the behaviour of rational functions at points of the curve.

3.1 Congruence

On the unit circle polynomials that differ by multiples of $(x^2 + y^2 - 1)$ have the same values. So as far as the circle is concerned, they are the same. We shall make a definition of congruence modulo $f(x, y)$ that reflects this idea for the curve C: $f(x, y) = 0$. The different possible polynomial functions on the affine curve will then correspond to the congruence classes modulo $f(x, y)$.

DEFINITION Let $f(x, y)$ be an irreducible polynomial in $K[x, y]$. Then we define polynomials $g(x, y)$ and $h(x, y)$ to be *congruent* modulo $f(x, y)$,

$$g(x, y) \equiv h(x, y) \pmod{f(x, y)}$$

if

$$g(x, y) - h(x, y) = q(x, y)f(x, y)$$

for some polynomial $q(x, y)$. The set of all polynomials congruent to a given polynomial $g(x, y)$ modulo $f(x, y)$ is called a *congruence class* modulo $f(x, y)$.

With polynomials in a single indeterminate, division with remainder enables us to pick out a 'best' member of each class, namely the common remainder on division by $f(x)$. For polynomials in two indeterminates, such a choice is no longer possible. So we shall have to operate with the classes themselves.

3.2 The coordinate ring

Addition and multiplication of polynomial functions on a curve make sense, so we shall introduce addition and multiplication of congruence classes. When you read 'add congruence class A to congruence class B' you should consider this as shorthand for 'take any polynomials $g \in A$ and $h \in B$ and add them; consider only the class of the result $g + h$'. This is a bit like the childhood rule 'odd + odd = even', if we take odd to mean the class of odd integers and even to mean the class of even integers.

DEFINITION Let $f(x, y)$ be an irreducible polynomial in $K[x, y]$. We define the *residue class ring* $K[x, y]/f(x, y)$ to have as its elements the congruence classes modulo $f(x, y)$. Addition and multiplication are defined by the following rule.

If A and B are congruence classes modulo f, choose $g \in A$ and $h \in B$ and define $A+B$ and AB to be the classes containing $g+h$ and gh respectively.

If C is the affine curve C: $f(x,y)=0$, then $K[x,y]/f(x,y)$ is called the *coordinate ring* of C and denoted by $K[C]$.

PROPOSITION *The definition above makes $K[x,y]/f(x,y)$ into an integral domain.*

Proof The main effort of the proof is expended in showing that the operations are well-defined. That means that the result of an operation is independent of the elements that are chosen in each class. The verification of the axioms then follows a routine course.

Addition and multiplication are indeed well-defined. For suppose that g_1 and g_2 both lie in A and h_1 and h_2 both lie in B. We must show that g_1+h_1 and g_2+h_2 lie in the same class, and also that g_1h_1 and g_2h_2 lie in the same class. By assumption f divides g_1-g_2 and h_1-h_2. Hence it divides $g_1+h_1-(g_2+h_2)$. So g_1+h_1 and g_2+h_2 lie in the same class. Also f divides $(g_1-g_2)h_1+g_2(h_1-h_2)=g_1h_1-g_2h_2$. So g_1h_1 and g_2h_2 lie in the same class.

To establish those axioms that express the identity of two formulae, we need only choose representatives of the classes in question and appeal to the validity of the formulae in $K[x,y]$. Thus to prove

A1 $$A+(B+C)=(A+B)+C,$$

we choose $g \in A$, $h \in B$, and $k \in C$. Then $g+(h+k)=(g+h)+k$, establishing the formula. This argument can be adapted to prove all the commutative, associative, and distributive laws.

The 0 and 1 classes are the classes containing the 0 and 1 of $K[x,y]$. Thus the 0 class is the set of multiples of $f(x,y)$.

The negative of the class A is the class containing the negatives of its elements (if you prefer: choose $g \in A$ and take the class of $-g$).

There remains the question of the cancellation law. That is proved using Proposition 2.3. Suppose that AB is the 0 class. That means that for $g \in A$ and $h \in B$, f divides gh. Since f is irreducible, it follows that f divides one of g or h. Thus $A=0$ or $B=0$. □

3.3 A convention

From now on we shall frequently be performing calculations with polynomials in the coordinate ring of a curve, but occasionally we shall wish to calculate in the full polynomial ring. To distinguish these cases we introduce the following convention.

CONVENTION Let K be a field; we shall denote polynomial rings over K by $K[\dot{x}]$, $K[\dot{x},\dot{y}]$ and their elements by $f(\dot{x})$ or $f(\dot{x},\dot{y})$ with dots over the indeterminates. The letters for the indeterminates are not restricted to x and y. Thus we may also use $f(\dot{u},\dot{v})$.

When the curve C: $f(x,y)=0$ is clear from the context we shall denote by $g(x,y)$ the element of its coordinate ring $K[C]=K[\dot{x},\dot{y}]/f(\dot{x},\dot{y})$ corresponding to the polynomial $g(\dot{x},\dot{y})$. Thus $g(x,y)=0$ states that $g(\dot{x},\dot{y}) \equiv 0 \pmod{f(\dot{x},\dot{y})}$.

3.4 The function field

The structure of the residue class ring $K[\dot{x}, \dot{y}]/f(\dot{x}, \dot{y})$ is closely related to the nature of the affine curve C: $f(x, y) = 0$. Indeed it is so closely linked that the structure can change under projective transformations. For example, consider the polynomial x. For any affine curve C_1: $f(x, y) = 0$, x is a polynomial function, but for another component of the same curve C_2: $g(u, v) = 0$, x corresponds to the function $1/v$. This function will usually not be equivalent to a polynomial.

To avoid having to check which polynomials remain polynomials under projective transformations, we extend our residue class ring to rational functions. These are invariant under all the transformations we need and are also the functions used to define geometric codes.

DEFINITION The function field $K(C)$ of the curve C: $f(x, y) = 0$ over the field K is the field of fractions of $K[C] = K[\dot{x}, \dot{y}]/f(\dot{x}, \dot{y})$. If $K(C)$ is isomorphic to the field of rational functions $f(\dot{t})/g(\dot{t})$ in some single indeterminate, then the curve is called *rational*.

Recall that the field of fractions of an integral domain D is constructed by taking fractions a/b, with $a, b \in D$ and $b \neq 0$. Fractions a/b and c/d are considered equal if $ad = bc$, and we use the usual rules of addition and multiplication:

$$\frac{a}{b} + \frac{c}{d} = \frac{(ad + bc)}{bd}; \qquad \frac{a}{b}\frac{c}{d} = \frac{ac}{bd}.$$

It is a straightforward exercise to show that, as the cancellation law holds in D, this construction produces a field with D isomorphic to the subring of fractions of the form $a/1$ (see Exercise 1).

EXAMPLE Consider the cubics $x^3 = y$ and $x^3 + y^3 = 1$. By replacing every occurrence of y by x^3, every polynomial in $K[\dot{x}, \dot{y}]/(\dot{x}^3 - \dot{y})$ can be reduced to a polynomial in x alone. Thus $K[\dot{x}, \dot{y}]/(\dot{x}^3 - \dot{y})$ is isomorphic to $K[\dot{x}]$ and so the curve is an example of a rational curve. An algebraic one-dimensional bug living on this curve could not tell that it was not on a straight line.

That phenomenon always occurs for quadratic curves (see Exercises 12-14), though in general the isomorphism does not take $K[C]$ to the polynomial ring $K[\dot{t}]$ (Exercise 12). By contrast cubic curves are not all rational. In particular, the second example is not rational (see Exercise 15). So by doing algebraic calculations a one-dimensional bug living on the curve $x^3 + y^3 = 1$ could determine that it was not on a straight line.

3.5 Independence of coordinate system

The function field $K(C)$ of a curve is the same for all three of its affine components. It does not depend on the choice of coordinate system.

PROPOSITION *Let $f(\dot{x}, \dot{y}) \neq a\dot{x}$ be an irreducible polynomial of degree d, and let $g(u, v) = \dot{v}^d f(1/\dot{v}, \dot{u}/\dot{v})$. Then the map σ: $h(\dot{x}, \dot{y}) \mapsto h(1/\dot{v}, \dot{u}/\dot{v})$ induces an isomorphism of the function fields of the curves $f(x, y) = 0$ and $g(u, v) = 0$.*

Notice that, unlike the defining polynomial of the curve, h is not multiplied by a power of v and that the transformation is defined independently of the curve in question.

EXAMPLE Consider the curve C: $x^3 + y^3 - 1 = 0$ and the two functions $\varphi_1 = x^3/(x+y)$ and $\varphi_2 = y^3/(x+y)$ in $K(C)$. Then

$$\varphi_1 + \varphi_2 = (x^3 + y^3)/(x+y) = 1/(x+y).$$

In the (u,v)-system C is defined by $1 + u^3 - v^3 = 0$.

$$\begin{aligned}\sigma(\varphi_1) &= (1/v^3)/(1/v + u/v) = 1/(v^2(1+u)),\\ \sigma(\varphi_2) &= u^3/(v^2(1+u)),\\ \sigma(1/(x+y)) &= v/(1+u).\\ \sigma(\varphi_1) + \sigma(\varphi_2) &= (1+u^3)/(v^2(1+u)) = v^3/(v^2(1+u))\\ &= v/(1+u) = \sigma(1/(x+y)).\end{aligned}$$

Proof Observe that $g(\dot{u}, \dot{v}) \neq av$ and that by Proposition 2.8 $g(\dot{u}, \dot{v})$ is absolutely irreducible. For the moment, consider the curves C: $f(x,y) = 0$ and D: $g(u,v) = 0$ to be different. We first consider σ to define a map from $K(\dot{x}, \dot{y})$ into $K(\dot{u}, \dot{v})$. This is an isomorphism with inverse map τ: $k(\dot{u}, \dot{v}) \mapsto k(\dot{y}/\dot{x}, 1/\dot{x})$. For later use we remark that $\tau(g(\dot{u}, \dot{v})/\dot{v}^d) = f(\dot{x}, \dot{y})$.

We restrict σ to polynomials $h(\dot{x}, \dot{y}) \in K[\dot{x}, \dot{y}]$ and note that if h has degree e the denominator of $\sigma(h(\dot{x}, \dot{y}))$ is v^e which is not divisible by $g(\dot{u}, \dot{v})$. Hence $\sigma(h)$ represents an element of $K(D)$. We shall show that $\sigma(h)$ represents 0 in $K(D)$ if and only if h is a multiple of f.

Since $\sigma(f(\dot{x}, \dot{y})) = g(\dot{u}, \dot{v})/\dot{v}^d$, multiples of $f(\dot{x}, \dot{y})$ are certainly mapped to functions representing 0. Suppose that $\sigma(h) = 0$ and let $\deg(h(\dot{x}, \dot{y})) = e$. As each term $b\dot{x}^r\dot{y}^s$ of h corresponds to a term $b\dot{u}^s/\dot{v}^{r+s}$ in $s(h)$, $\dot{v}^e h(1/\dot{v}, \dot{u}/\dot{v})$ is a polynomial of degree $\leq e$ that is divisible by $g(\dot{u}, \dot{v})$, say $\dot{v}^e h(1/\dot{v}, \dot{u}/\dot{v}) = g(\dot{u}, \dot{v})k(\dot{u}, \dot{v})$ in $K[\dot{u}, \dot{v}]$. Then $\deg(k(\dot{u}, \dot{v})) \leq e - d$ (because $\deg(g(\dot{u}, \dot{v})) = d$) and

$$(\sigma h)(\dot{u}, \dot{v}) = (g(\dot{u}, \dot{v})/\dot{v}^d)(k(\dot{u}, \dot{v})/\dot{v}^{e-d}).$$

Hence

$$h(\dot{x}, \dot{y}) = \tau\sigma h(\dot{x}, \dot{y}) = f(\dot{x}, \dot{y})\tau k(\dot{x}, \dot{y})\dot{x}^{d-e}.$$

Since $d - e \geq \deg k$ it follows that $\tau k(\dot{x}, \dot{y})\dot{x}^{d-e}$ is a polynomial, and so $f(\dot{x}, \dot{y})$ divides $h(\dot{x}, \dot{y})$.

Thus $\sigma(h_1) = \sigma(h_2)$ in $K(D)$ if and only if h_1 and h_2 are congruent modulo $f(x,y)$. So σ defines a map from $K[C]$ into $K(D)$ for which the only element of $K[C]$ that goes to 0 is 0 itself. It is obvious that $\sigma(h_1 + h_2) = \sigma(h_1) + s(h_2)$ and $\sigma(h_1 h_2) = \sigma(h_1)\sigma(h_2)$. We extend the map to the whole of $K(C)$ by defining $\sigma(h_1/h_2) = \sigma(h_1)/\sigma(h_2)$ when $h_2 \neq 0$ in $K[C]$. That produces an embedding of $K(C)$ in $K(D)$.

Reversing the roles of f and g, the same argument shows that τ induces an embedding of $K(D)$ in $K(C)$. But σ and τ are inverse maps. Thus the two fields are isomorphic. □

3.6 Evaluating functions at points

A polynomial $f(x, y) \in K[C]$ can be evaluated at all (x, y)-points of C. Obviously, if the point $P: (x{=}\alpha, y{=}\beta)$ has coefficients in an extension field E of K the value of f may be in E. However, a rational function φ may not be defined at all for certain points, for instance $1/x$ is not defined at the origin.

EXAMPLE Consider the function $f(x, y) = y/(x + 1)$ defined on the real unit circle $x^2 + y^2 = 1$. In the (x, y)-system the only point on the circle with $x = -1$ is $(-1, 0)$. So this function is defined for all (x, y)-points except $(-1, 0)$. What about (u, v)-points?. The equation of the circle in the (u, v)-system is $v^2 - u^2 = 1$ and our function becomes $(u/v)/((1/v) + 1) = u/(v + 1)$. The only point on the circle with $v = -1$ is $(u{=}0, v{=}{-1})$, which is the same as our original point. The only point that has not yet been considered is $(w{=}0, z{=}0)$, but that does not lie on the circle. So the function is defined for all points except possibly one.

If the numerator of the function were non-zero at $(-1, 0)$, then it would be clear that the function is not defined at that point. However, the numerator is also 0 at $(-1, 0)$. So we must use the equation of the circle to transform the function. Using the substitution $y^2 = 1 - x^2$ we find that

$$f(x, y)^2 = (1 - x^2)/(1 + x)^2 = (1 - x)/(1 + x).$$

Thus $f = \sqrt{1 - x}/\sqrt{1 + x}$. The denominator of this function is zero at $(-1, 0)$ and the numerator is not. That establishes that $f(x, y)$ is not defined at $(-1, 0)$. It is defined at all other points.

Now consider the function $1/f = (x + 1)/y$. The argument we have just given shows that $1/f$ is defined at $(-1, 0)$ and it is also defined at all (x, y)-points on the circle except $(1, 0)$. In the (u, v)-system $1/f = (v + 1)/u$ which is certainly defined for all points except $(u{=}0, v{=}\pm 1)$. These points are the same as $(x{=}\pm 1, y{=}0)$, which have already been considered. This function is defined everywhere except at $(1, 0)$.

Finally consider the function $f + 1/f$. Using the equation of the circle we can simplify this:

$$f + 1/f = \frac{y}{x + 1} + \frac{x + 1}{y} = \frac{y^2 + x^2 + 2x + 1}{y(x + 1)} = \frac{2x + 2}{y(x + 1)} = \frac{2}{y}.$$

That function is defined everywhere except $(x{=}\pm 1, y{=}0)$. In the (u, v)-system it becomes $2v/u$ which is defined everywhere except at the points we have already found.

This discussion leads to the following definition.

DEFINITION Let $P: (x{=}\alpha, y{=}\beta)$ be a point. The ring $K[\dot{x}, \dot{y}]_P$ is the set of all rational functions $\varphi = g(\dot{x}, \dot{y})/h(\dot{x}, \dot{y})$ such that $h(\alpha, \beta) \neq 0$.

If P lies on a curve $C: f(x, y) = 0$ the *local ring* $K[C]_P$ is the set of all functions in $K(C)$ that can be represented by rational functions in $K[\dot{x}, \dot{y}]_P$.

The local ring of a point does not depend on the coordinate system used for its representation.

PROPOSITION *Let* $P\colon (x{=}\alpha, y{=}\beta) = Q\colon (u{=}\gamma, v{=}\delta)$; *then* $K[C]_P = K[C]_Q$.

Proof It will suffice to show that $K[C]_P \subseteq K[C]_Q$ for then the result follows by symmetry. So let $g(x,y)/h(x,y) \in K[C]_P$ where the representatives have been chosen so that $h(\alpha,\beta) \neq 0$. Set $\deg g = d$ and $\deg h = e$. Then

$$\frac{g(x,y)}{h(x,y)} = \frac{g(1/v,u/v)}{h(1/v,u/v)} = \frac{v^e(v^d g(1/v,u/v))}{v^d(v^e h(1/v,u/v))}.$$

Both $v^d g(1/v,u/v)$ and $v^e h(1/v,u/v)$ lie in $K[u,v]$ and as $P\colon (\alpha,\beta) = Q\colon (\gamma,\delta)$ we must have $\delta = 1/\alpha$ which cannot be 0. Thus the denominator of the right hand fraction has value $\delta^{d+e}h(\alpha,\beta)$ which is non-zero. That proves that $g(x,y)/h(x,y) \in K[C]_Q$. □

3.7 Shifting to the origin

When dealing with local properties of functions it is often convenient to move the point P we are considering to the origin. To that end we define the polynomial $f_{\alpha,\beta}(\dot{x},\dot{y})$ by the condition that

$$f_{\alpha,\beta}(\dot{x},\dot{y}) = f(\dot{x}-\alpha,\dot{y}-\beta).$$

The following proposition shows that $f_{\alpha\, b}(\dot{x},\dot{y})$ shares all the essential algebraic properties of $f(\dot{x},\dot{y})$.

PROPOSITION *Let* $f(\dot{x},\dot{y})$ *and* $g(\dot{x},\dot{y})$ *be polynomials. Then*
(a) $f_{\alpha,\beta} + g_{\alpha,\beta} = (f+g)_{\alpha,\beta}(\dot{x},\dot{y})$ *and* $f_{\alpha,\beta}g_{\alpha,\beta} = (fg)_{\alpha,\beta}(\dot{x},\dot{y})$;
(b) $f_{\alpha+\gamma,\beta+\delta}(\dot{x},\dot{y}) = (f_{\alpha,\beta})_{\gamma,\delta}(\dot{x},\dot{y})$;
(c) $\deg f_{\alpha,\beta}(\dot{x},\dot{y}) = \deg f(\dot{x},\dot{y})$; *and*
(d) $f_{\alpha\beta}(\dot{x},\dot{y})$ *is (absolutely) irreducible if and only if* $f(\dot{x},\dot{y})$ *is (absolutely) irreducible.*

Proof Statements (a) and (b) follow directly from the definition.

(c) For a monomial $f(\dot{x},\dot{y}) = \dot{x}^m\dot{y}^n$, $f_{\alpha,\beta}(\dot{x},\dot{y}) = (\dot{x}-\alpha)^m(\dot{y}-\beta)^n$ which has degree $m+n$ by the formula for the degree of a product. Regarding a general polynomial $f(\dot{x},\dot{y})$ as a sum of monomials, it follows that $f_{\alpha,\beta}(\dot{x},\dot{y})$ is a sum of polynomials of degree at most equal to $\deg f(\dot{x},\dot{y})$. Hence $\deg f_{\alpha,\beta}(\dot{x},\dot{y}) \leq \deg f(\dot{x},\dot{y})$. By statement (b) $f(\dot{x},\dot{y}) = f_{0,0}(\dot{x},\dot{y}) = (f_{\alpha,\beta})_{-\alpha,-\beta}(\dot{x},\dot{y})$. So by symmetry $\deg f(\dot{x},\dot{y}) \leq \deg f_{\alpha,\beta}(x,y)$.

(d) By statement (a) $f(\dot{x},\dot{y}) = g(\dot{x},y)h(\dot{x},\dot{y})$ holds if and only if we have $f_{\alpha,\beta}(\dot{x},\dot{y}) = g_{\alpha,\beta}(\dot{x},\dot{y})h_{\alpha,\beta}(\dot{x},\dot{y})$. Thus $f(\dot{x},\dot{y})$ has a non-trivial factorization if and only if $f_{\alpha,\beta}$ has a non-trivial factorization. □

As $f_{\alpha,\beta}(0,0) = f(\alpha,\beta)$ we can shift a given point of interest to the origin. We shall make use of this process in several proofs.

3.8 Divisibility by $x - \alpha$

For later use we need to establish that elements of the local ring $K[C]_P$ of a point $P\colon (\alpha, \beta)$ cannot be represented as multiples of arbitrarily high powers of $x - \alpha$. That fact, proved in the proposition below, is a special case of the Hilbert basis theorem.

PROPOSITION *Let $P\colon (\alpha, \beta)$ lie on the curve $C\colon f(x, y) = 0$ and let $\varphi(x, y) \neq 0$ be a function in $K[C]_P$. Then there are only finitely many natural numbers n such that $\varphi(x, y)$ can be written in the form $(x - \alpha)^n \psi(x, y)$ with $\psi(x, y) \in K[C]_P$.*

Proof Shift to the origin and assume that $P = (0, 0)$. If $\varphi = g/h$ is an element of $K[C]_P$ for which the statement of the proposition is false, then the statement also fails for g. For if $\varphi = x^n \psi$, then $g = x^n h\psi$, and $\psi \in K[C]_P$ implies $h\psi \in K[C]_P$. So we may assume that $\varphi = g \in K[C]$.

(1) Let I_n be the set of polynomials $h(\dot{x}, \dot{y}) \in K[\dot{x}, \dot{y}]$ such that in $K[C]_P$ we have

$$gk(x, y) = x^n h(x, y), \qquad (*)$$

for some polynomial $k(x, y)$ depending on h. Since we can multiply eqn $(*)$ by x it is clear that $I_n \subseteq I_{n+1}$. Furthermore if $I_n = I_{n+1}$, then $g = x^m \psi$ with $\psi \in K[C]_P$ implies that $m \leq n$. For in that case any h in I_{n+1} satisfies eqn $(*)$ and so

$$g = x^{n+1} h(x, y)/k_1(x, y)$$

in $K(C)$ implies that $gk_1(x, y) = xgk(x, y)$ and hence $k_1(x, y) = xk(x, y)$. Therefore $k_1(0, 0) = 0$. Thus there is no rational function $\psi \in K[C]_P$ with $x^{n+1}\psi = g$. *A fortiori* there can be no rational function $\psi \in K[C]_P$ with $x^m \psi = g$ for $m > n + 1$.

(2) Now consider the set $J = \bigcup_{n=0}^{\infty} I_n$ and write all the polynomials $h(\dot{x}, \dot{y}) \in J$ in the form

$$a_0(\dot{x}) + a_1(\dot{x})\dot{y} + \cdots + a_r(\dot{x})\dot{y}^r$$

with 'highest coefficient' $a_r(\dot{x}) \neq 0$ (where of course r varies with h). Let A_r be the set of all polynomials $a_r(\dot{x})$ that occur as highest coefficients of polynomials in J of degree exactly r in $\dot{y}$. For each A_r we choose a polynomial $a_r(\dot{x}) \in A_r$ of smallest possible degree in $\dot{x}$. Since we can multiply polynomials in J by $\dot{x}$ it is clear that $A_r \subseteq A_{r+1}$. Hence $\deg a_{r+1} \leq \deg a_r$. Since these degrees cannot decrease indefinitely, there must be an r such that $\deg a_r = \deg a_{r+1}$.

(3) We shall show that every element of A_r is a multiple of a_r. For if $b(\dot{x}) \in A_r$, let

$$b(\dot{x}) = q(\dot{x})a_r(\dot{x}) + c(\dot{x})$$

with $\deg c(\dot{x}) < \deg a_r(x)$. Choose polynomials $h_1(\dot{x}, \dot{y})$, $h_2(\dot{x}, \dot{y})$ in J of degree r in $\dot{y}$ with leading coefficients $a_r(\dot{x})$ and $b(x)$. Then h_1 and h_2 satisfy equations $gk_1(x, y) = x^m h_1(x, y)$ and $gk_2(x, y) = x^n h_2(x, y)$. We may assume that $m = n$ by multiplying one of the equations by a power of x. It follows that

$$g(k_2(x, y) - q(x)k_1(x, y)) = x^n(h_2(x, y) - q(x)h_1(x, y)).$$

Therefore $h_2(\dot{x}, \dot{y}) - q(\dot{x})h_1(\dot{x}, \dot{y}) \in J$ but the coefficient of $\dot{y}^r$ in this polynomial is $b(\dot{x}) - q(\dot{x})a_r(\dot{x}) = c(\dot{x})$ and its degree in $\dot{y}$ is at most r. Hence if $c(\dot{x}) \neq 0$ it lies in A_r. But the degree of $c(x)$ in $\dot{x}$ is less than that of $a_r(x)$, which is the smallest in A_r. So $c(\dot{x})$ must be 0.

(4) From the fact that $A_r \subseteq A_{r+1}$ it now follows that a_r is a multiple of a_{r+1}. Thus if they have the same degree they differ by multiplication by a non-zero constant. That implies that they have the same multiples and so $A_r = A_{r+1}$.

Fix r such that $A_{r+1} = A_r$. For $s = 0, \ldots, r$ choose a polynomial $h_s(\dot{x}, \dot{y}) \in J$ of degree s in $\dot{y}$ with leading coefficient $a_s(\dot{x})$. There exists an n such that all these $r+1$ polynomials lie in I_n. For $s > r$ we put $h_s(\dot{x}, \dot{y}) = x^{s-r} h_r(x, y)$, which also lies in I_n.

(5) We now show by induction on the degree in $\dot{y}$ that $I_{n+1} = I_n$. Let $h(\dot{x}, \dot{y})$ lie in I_{n+1} and have degree s in y and let the leading term of h be $q(\dot{x})a_s(\dot{x})$. Let

$$gk(x, y) = x^{n+1} h(x, y) \quad \text{and} \quad gk_s(x, y) = x^n h_s(x, y).$$

Then

$$g(x, y)(k(x, y) - xq(x)k_s(x, y)) = x^{n+1}(h(x, y) - q(x)h_s(x, y)).$$

So $h(\dot{x}, \dot{y}) - q(\dot{x})h_s(\dot{x}, \dot{y}) \in I_{n+1}$. This has degree less than s in $\dot{y}$. If $s = 0$ the polynomial is zero (which trivially lies in I_n), otherwise we can assume by the induction hypothesis that it lies in I_n.

Thus for some polynomial $k_t(x, y)$ we can write $gk_t = x^n(h - qh_s)$ and so

$$\begin{aligned} g(x, y)(k_t(x, y) + q(x)k_s(x, y)) &= x^n(h(x, y) - q(x)h_s(x, y)) + q(x)x^n h_s(x, y) \\ &= x^n h(x, y) \end{aligned}$$

showing that $h(\dot{x}, \dot{y}) \in I_n$. It follows from step (1) that $g = x^m \psi$ with $\psi \in K[C]_P$ implies that $m \leq n$. □

3.9 Simple points

In Example 3.6 we manipulated the function $y/(x+1)$ on the unit circle to determine whether it was defined at the point $(-1, 0)$. No justification was given why such a manipulation should always be possible, because, in fact, sometimes it is not. Whether the manipulation is possible for all functions depends on the properties of the point on the curve. The definition below gives the name *simple* to well-behaved points. The next few paragraphs are devoted to a detailed examination of simple points, culminating in Paragraph 3.12 with an easy test to determine whether a point is simple.

DEFINITION P is called *simple*, or *non-singular,* if for any rational function φ at least one of φ and $1/\varphi$ lies in $K[C]_P$.

A simple point P is a *pole* of a rational function φ if φ does not lie in $K[C]_P$; it is a *zero* of φ if $1/\varphi$ does not lie in $K[C]_P$.

Similar definitions hold for (u, v)-points and (w, z)-points.

A point which is not simple is called *singular.*

EXAMPLES Some examples of curves over R will elucidate this idea.

Consider the parabola $C\colon y = x^2$ and the point $P\colon (0, 0)$. Given any rational function g/h we can replace all occurrences of y by x^2 and so assume that g and h are both polynomials in x alone. Cancelling any common factors we can also ensure that x does not divide both g and h. If x does not divide h, then g/h is defined at P, and if it does not divide g, h/g is defined at P. For this curve $P\colon (0, 0)$ is a simple point.

Now consider the cubic $y^2 = x^3$. If you draw this curve you will see that it has a sharp peak at the origin $P\colon (0, 0)$ with a horizontal tangent, a so-called *cusp*. The function y/x has as its square $y^2/x^2 = x^3/x^2 = x$. So its 'value' at P should be 0. Nevertheless it is not possible to write $y/x = g(x, y)/h(x, y)$ with $h(x, y) \neq 0$. For then

$$\dot{y}h(\dot{x}, \dot{y}) = \dot{x}g(\dot{x}, \dot{y}) + k(\dot{x}, \dot{y})(\dot{y}^2 - \dot{x}^3)$$

for some polynomial $k(\dot{x}, \dot{y})$. By assumption $h(\dot{x}, \dot{y})$ has a non-zero constant term a and so the left hand side contains a term $a\dot{y}$. However, every term on the right hand side involves $\dot{x}$ or $\dot{y}^2$. So the two sides cannot be equal. This is an example of a singular point.

Finally consider the α-shaped curve $y^2 = x^3 + x^2$. This crosses itself at $P\colon (0, 0)$ and has two tangents there with slopes ± 1. Depending on the branch one approaches P on, the value of y/x at P tends to ± 1, so it is not possible to assign a definite value to y/x. The same argument as for the cusp shows that y/x cannot be written as a fraction $g(x, y)/h(x, y)$ with $h(x, y) \neq 0$. This is also a singular point.

3.10 Order functions

At a simple point P on a curve we can refine the notion of poles and zeros by defining an order ν for functions $\varphi \in K(C)$ such that $\nu(\varphi) > 0$ if P is a zero of φ and $\nu(\varphi) < 0$ if P is a pole of φ. This mimics the definition of the order of a complex rational function at a point in the complex plane.

EXAMPLE Let $K = \mathsf{R}$ and consider the circle $C\colon x^2 + y^2 = 1$, including the points with coordinates in C.

Let $\varphi = x(x-1)^2/(y-1)^2$. Then for $P\colon (1, 0)$, $\nu(\varphi) = 2$, because of the factor $(x-1)^2$ in the numerator. For $P\colon (0, 1)$, $\nu(\varphi) = -1$, because we have factors x in the numerator and $(y-1)^2$ in the denominator.

Let $\psi = x^2 + 1 = (x + \mathsf{i})(x - \mathsf{i})$ (where $\mathsf{i}^2 = -1$). Then for $P\colon (\mathsf{i}, -\mathsf{i})$, and $\bar{P}\colon (-\mathsf{i}, \mathsf{i})$, $\nu(\psi) = 1$. On the other hand, at these points $\nu(\varphi) = 0$ and this indicates a non-zero value $\pm\mathsf{i}$.

It is also possible for a function to have poles or zeros at (u, v)- or (w, z)-points. Consider the points $Q\colon (u{=}\mathsf{i}, v{=}0)$ and $\bar{Q}\colon (u{=}-\mathsf{i}, v{=}0)$ and the function $\chi = x$. In the (u, v)-system χ becomes $1/v$, so it has order -1 at Q and $\bar{Q}$. It will turn out that if the curve has no singularities, then all non-constant polynomials have some poles at the horizon.

The order of any real rational function is always the same at conjugate complex points P and $\bar{P}$, and if the order is 0, then the values at conjugate points are conjugate.

A natural way of defining an order function at $P\colon (\alpha, \beta)$ is to let the order of $0 \neq \varphi \in K[C]_P$ be the highest power of $x - \alpha$ dividing φ. We call this the x-order of φ. We can similarly define the y-order. There are several problems that we have to overcome to make this definition work.
(a) We would like to show that a function φ with x-order 0 has a non-zero value at P. That is not always true. We shall call the order *good* if this holds.
(b) For the definition to be worthwhile, we must know that at least one of the x- and y-orders at P is good. That is true precisely for simple points.
(c) We would like the x-order and y-order at P to be the same. That is true if both are good.

DEFINITION Let $C\colon f(x,y) = 0$ be a curve and let $P\colon (x{=}\alpha, y{=}\beta)$ be a point of C with $\alpha, \beta \in K$. Let $\varphi(x,y) \in K(C)$. If $\varphi = (x-\alpha)^n \psi$ with $\psi \in K[C]_P$ but there is no element $\psi' \in K[C]_P$ with $\varphi = (x-\alpha)^{n+1}\psi'$, then the *$x$-order* of φ, $\nu_{P,x}(\varphi)$, is defined to be n.

The x-order is called *good* if for $\varphi \in K[C]_P$, $\nu_{P,x}(\varphi) = 0$ implies that $\varphi(P) \neq 0$.

The *y-order* is defined analogously. I omit subscripts that are clear from the context.

It is clear that if $\nu_{P,x}(\varphi) = n$ then $n \geq 0$ if and only if $P \in K[C]_P$. Further, $n > 0$ implies that P is a zero of φ, but the converse may not hold.

EXAMPLE Consider the circle $x^2 + y^2 = 1$ and the point $P\colon (x{=}0, y{=}1)$. The x-order of x^n is obviously n. If the base field is R, then

$$\dot{y} - 1 \equiv -\dot{x}^2/(\dot{y}+1) \pmod{(\dot{x}^2 + \dot{y}^2 - 1)},$$

so the x-order of $y - 1$ is 2.

Similarly $\dot{y} \equiv (-\dot{x}^2 + \dot{y} + 1)/(\dot{y}+1)$ and $\dot{y} + 1 \equiv (-\dot{x}^2 + \dot{y} + 2)/(y+1)$ have x-order 0. That is consistent with the fact that these functions have non-zero values at P.

The x-order of $y^3 + y^2 - y - 1 = (y-1)(y+1)^2$ should be 2 because $y+1$ is non-zero at P, and indeed $y^3 + y^2 - y - 1 = x^2(x^2-2)/(y+1)$.

The y-order is not good. For the y-order of x is 0, because the equation

$$\dot{x} \equiv (\dot{y} - 1)g(\dot{x}, \dot{y}) + h(\dot{x}, \dot{y})(\dot{x}^2 + \dot{y}^2 - 1)$$

has no solutions, but nevertheless x has value 0 at P.

As you can see, the calculations for $P\colon (\alpha, \beta)$ depend on expanding polynomials in terms of $\dot{x} - \alpha$ and $\dot{y} - \beta$. We can always do that and the fact that P lies on $C\colon f(x,y) = 0$ is equivalent to the fact that in this expansion $f(\dot{x}, \dot{y})$ has no constant term.

3.11 Good orders and simple points

We can now provide a simple theoretical test to determine when the x-order or y-order at a point is good, and use it to show that the three desired conditions of the previous paragraph are satisfied for simple points..

THEOREM *Let C: $f(x,y)=0$ be a curve and let P: $(x{=}\alpha, y{=}\beta)$ be a point of C with $\alpha, \beta \in K$.*
(a) If $(y-\beta)/(x-\alpha) \in K[C]_P$, then the x-order is good.
(b) If P is simple then the x-order is good or the y-order is good.
(c) If the x-order is good then P is simple.
(d) If the x-order is good, then $(y-\beta)/(x-\alpha) \in K[C]_P$.
(e) If both the x-order and y-order are good, then they are the same.

Proof We shift to the origin and assume that $\alpha=\beta=0$.

(a) Suppose that $y/x \in K[C]_P$. We first show that $\nu_x(g)=0$ implies $g(0,0)\neq 0$ for a polynomial $g(x,y) \in K[C]$. For if $g(0,0)=0$ then $g(\dot{x},\dot{y})$ is congruent modulo $f(\dot{x},\dot{y})$ to a polynomial with no constant term. As $f(\dot{x},\dot{y})$ itself has no constant term, it follows that $g(\dot{x},\dot{y}) = \dot{x}g_1(\dot{x},\dot{y}) + \dot{y}g_2(\dot{x},\dot{y})$. Then in $K[C]_P$ we have

$$g(x,y) = x\left(g_1(x,y) + \frac{y}{x}g_2(x,y)\right),$$

showing that $\nu_x(g) \geq 1$.

It follows for $\varphi \in K[C]_P$ that $\nu_x(\varphi)=0$ implies that $\varphi(0,0)\neq 0$, for the denominator of φ has a non-zero value by assumption and we have just shown that the numerator of φ has a non-zero value. Thus the x-order is good.

(b) If P is simple, then one of y/x and x/y lies in $K[C]_P$. Part (a) tells us that in the first case the x-order is good. By analogy the y-order is good in the second case.

(c) Suppose that the x-order is good and we have an arbitrary rational function $\varphi = g(x,y)/h(x,y)$ in $K(C)$, with $\nu_x(g)=m$ and $\nu_x(h)=n\leq m$. Then $g(x,y) = x^m g_1(x,y)$ and $h(x,y)=x^n h_1(x,y)$, with $g_1(0,0), h_1(0,0)\neq 0$. Hence $\varphi = x^{m-n}g_1(x,y)/h_1(x,y)$ lies in $K[C]_P$. Similarly if $n>m$, then $1/\varphi \in K[C]_P$. So P is simple.

(d) As $y\in K[C]$ and y has value 0 at the origin, it follows from the goodness of the x-order that $\nu_x(y)>0$. Hence $y = x\varphi$ with φ in $K[C]_P$.

(e) We have that both x/y and y/x lie in $K[C]_P$. Thus if $\varphi = x^n\psi$ with $\psi \in K[C]_P$ we also have $\varphi = y^n(x/y)^n\psi$ and $(x/y)^n\psi \in K[C]_P$, so $\nu_x(\varphi)\leq\nu_y(\varphi)$. By symmetry it follows that $\nu_x(\varphi)=\nu_y(\varphi)$. □

3.12 A test for singularity

There is an easy practical test for singularity based on the formal partial derivatives of the polynomial defining the curve.

DEFINITION Let $f(\dot{x},\dot{y}) = \sum a_{ij}\dot{x}^i\dot{y}^j$ be a polynomial. Then $\partial f/\partial\dot{x}$ is defined by the formula $\sum i\cdot a_{ij}\dot{x}^{i-1}\dot{y}^j$ (here the factor i is an integer index).

THEOREM *Let C: $f(x,y)=0$ be an affine curve, and let P: $(x{=}\alpha, y{=}\beta)$ be a point of C.*
(a) *If $\partial f/\partial \dot{y}(\alpha,\beta) \neq 0$, then $(y-\beta)/(x-\alpha) \in K[C]_P$ and P is simple.*
(b) *If $(y-\beta)/(x-\alpha) \in K[C]_P$, then $\partial f/\partial \dot{y}(\alpha,\beta) \neq 0$.*
(c) *If $\partial f/\partial \dot{y}(\alpha,\beta) = 0$ and $\partial f/\partial \dot{x}(\alpha,\beta) = 0$, then P is singular.*

EXAMPLE Again taking the circle and the point P: $(0,1)$, then $f = \dot{x}^2 + \dot{y}^2 - 1$. Thus $\partial f/\partial \dot{y}(0,1) = 2$, so $(y-1)/x \in K[C]_P$ and the x-order is good. On the other hand $\partial f/\partial \dot{x}(0,1) = 0$ so $x/(y-1) \notin K[C]_P$ and the y-order is bad. These calculations confirm the results of the previous example.

Proof We shift to the origin so that $P = (0,0)$. Then

$$f(\dot{x},\dot{y}) = a\dot{x} + b\dot{y} + \dot{x}^2 f_1(\dot{x}) + \dot{y}^2 f_2(\dot{y}) + \dot{x}\dot{y} f_3(\dot{x},\dot{y}), \qquad (*)$$

and $a = \partial f/\partial x(0,0)$ and $b = \partial f/\partial y(0,0)$. There is no constant term, because P: $(0,0)$ lies on the curve.

(a) Assume $b \neq 0$, gather terms involving y, and write

$$\dot{y}(b + \dot{y} f_2(\dot{y}) + \dot{x} f_3(\dot{x},\dot{y})) = f(\dot{x},\dot{y}) - \dot{x}(a + \dot{x} f_1(\dot{x})).$$

Then in $K(C)$ we have

$$\frac{y}{x} = \frac{-(a + x f_1(x))}{b + y f_2(y) + x f_3(x,y)},$$

which lies in $K[C]_P$ because the denominator on the right is non-zero at the origin.

(b) Suppose conversely that $y/x = g(x,y)/h(x,y)$ with $h(0,0) = c \neq 0$. Multiplying up and working in $K[\dot{x},\dot{y}]$ we obtain

$$\dot{y} h(\dot{x},\dot{y}) = \dot{x} g(\dot{x},\dot{y}) + k(\dot{x},\dot{y}) f(\dot{x},\dot{y})$$

for some polynomial $k(\dot{x},\dot{y})$. The left hand side of this equation contains a non-zero term $c\dot{y}$ which must also appear on the right hand side. It cannot appear in $\dot{x} g(\dot{x},\dot{y})$. So it must appear in $k(\dot{x},\dot{y}) f(\dot{x},\dot{y})$. From eqn $(*)$ this can only be in the form $k_0 b y$, where k_0 is the constant term of $k(\dot{x},\dot{y})$. Hence $b \neq 0$ as required.

(c) If both $\partial f/\partial \dot{x}$ and $\partial f/\partial \dot{y}$ are zero at P, then part (b) implies that neither y/x nor x/y lies in $K[C]_P$ and so P is singular. □

3.13 Smooth curves

Let C be the curve $f(x,y) = 0$, defined over a field K. Theorem 3.12 tells us that a point $(x{=}\alpha, y{=}\beta)$ with coefficients in an extension field E of K is simple if and only if $\partial f/\partial x(\alpha,\beta) \neq 0$ or $\partial f/\partial y(\alpha,\beta) \neq 0$. For points defined in the (u,v)- and (w,z)-systems analogous statements hold. A corollary of the zeros theorem of the next chapter is that a curve has only finitely many singular points.

DEFINITION A curve C with only simple points is called *smooth*. If all its (x, y)-points are simple it is called (x, y)-*smooth*.

The *order function* ν_P at a simple point P is whichever of x-order and y-order is good.

EXAMPLES Calculations involving singularity depend strongly on the characteristic of the underlying field and tend to be easier for finite fields than for the classical fields.

The cubic curve C: $x^3 + y^3 = 1$. The curve is smooth except over fields of characteristic 3, because $\partial f/\partial x = 3x^2$ and $\partial f/\partial y = 3y^2$ and $(x{=}0, y{=}0)$ is not a point of C. The same argument works in the other coordinate systems.

Consider the point P: $(x{=}0, y{=}1)$. As $\partial f/\partial x = 0$ and $\partial f/\partial y = 3 \cdot 1 \neq 0$ at P, we must use the x-order. So the order of x is 1. To calculate the order of $y - 1$, rewrite $f(x, y) = x^3 + (y-1)(1+y+y^2)$. Thus $y - 1 = x^3/(1+y+y^2)$ and hence the order of $y - 1$ is 3.

The Klein quartic $x^3y + y^3 + x = 0$. Over a field of characteristic 2 the partial derivative $\partial f/\partial x$ is $x^2y + 1$. If this is 0, then $x^2y = 1$ and $x^3y + y^3 + x = x + y^3 + x = y^3$, so $y = 0$, but then $x = 0$ and $x^2y + 1 = 1$, giving a contradiction. A similar argument holds in the other coordinate systems. Thus the curve is smooth for fields of characteristic 2.

Consider the point P: $(x{=}0, y{=}0)$. As $\partial f/\partial y(0, 0) = 0$, we must use the y-order. Writing $f(x, y)$ as $x(x^2y+1)+y^3$ we see that $x = y^3/(x^2y+1)$. So the y-order of x is 3, while the order of y is 1.

The quintic curve C: $x^5 + y^5 = 1$. This curve has $\partial f/\partial x = 5x^4$ and $\partial f/\partial y = 5y^4$. As $(x{=}0, y{=}0)$ is not on the curve, it is smooth except over fields of characteristic 5. Again, the same argument works in the other coordinate systems.

The order function at P: $(x{=}0, y{=}1)$ can be obtained in the same way as for $x^3 + y^3 = 1$. We write $f(x, y) = (y-1)(y^5 + y^4 + y^2 + y + 1) + x^5$, and see that $\nu(y - 1) = 5$, while $\nu(x) = 1$.

3.14 Conjugate points

Notice that in Examples 2.10–2.12, the points in the extension fields are grouped together by semi-colons. The coordinates of points in the same group have the same minimal polynomials. Hence, if a rational function defined over F_2 is zero on one point of a group it is zero on all the others as well. That means that as far as F_2 is concerned we cannot distinguish which of the points in a group we have picked out. That is rather like the fact that as far as real functions are concerned we cannot distinguish between the complex conjugates $+\mathrm{i}$ and $-\mathrm{i}$.

DEFINITION Two points $(x{=}\alpha, y{=}\beta)$ and $(x{=}\alpha', y{=}\beta')$ with coefficients in an extension field E of K are called *conjugate* if $g(\alpha, \beta) = 0 \iff g(\alpha', \beta') = 0$ for all polynomials $g(\dot{x}, \dot{y})$ in $K[\dot{x}, \dot{y}]$.

EXAMPLE We continue with Example 2.10, the cubic curve $x^3 + y^3 + 1$ over F_2, and determine the conjugacy classes of the points in fields of order up to 8.

Distinct points over F_2 are not conjugate.

Consider the points $(10,0)$ and $(11,0)$ over F_4. Suppose a polynomial $g(\dot{x},\dot{y}) \in \mathsf{F}_2[\dot{x},\dot{y}]$ is zero at $(10,0)$. Substituting $\dot{y}=0$ leaves a polynomial $g(\dot{x},0)$ in x alone. This must be zero for $\dot{x}=10$. The minimal polynomial of 10 over F_2 is $\dot{x}^2+\dot{x}+1$ and so $\dot{x}^2+\dot{x}+1$ divides $g(\dot{x},0)$. But 11 is also a root of $\dot{x}^2+\dot{x}+1$ and hence $g(11,0)=0$. The same argument works in reverse and thus these two points are conjugate.

On the other hand $(10,0)$ and $(0,10)$ are not conjugate, because $x^2+\dot{x}+1$ has value 0 at $(10,0)$ and value 1 at $(0,10)$.

Now consider the points $(2',5')$ and $(4',6')$ over F_8. In that field $4'=2'^2$ and $6'=5'^2$. So if $g(\dot{x},\dot{y}) \in \mathsf{F}_2[\dot{x},\dot{y}]$, then $g(4',6')=g(2',5')^2$ because squaring preserves addition in characteristic 2 and the coefficients of g are unchanged by squaring. The same argument works also for $(2',5')$ and $(7',3')=(2'^4,5'^4)$.

On the other hand $\dot{x}^3+\dot{x}^2+1$ has a zero at $2'$ but its value at $5'$ is $3' \neq 0$. Thus $(2',5')$ and $(5',2')$ are not conjugate.

In Paragraph 3.6 we observed that the order functions of complex conjugate points on real polynomials were the same. From the real point of view they are identical twins. That holds in the more general situation of algebraic curves.

PROPOSITION (a) *Let P and P' be conjugate points over a field F; then if one lies on a curve C defined over F, then so does the other. If one is simple, then so is the other.*

(b) *Conjugate simple points of a curve C define the same order function.*

Proof (a) If C has the equation $f(x,y)=0$, and the points $P\colon(x{=}\alpha,y{=}\beta)$ and $P'\colon(x{=}\alpha',y{=}\beta')$ are conjugate, then the fact that P lies on C implies $f(\alpha,\beta)=0$ and that implies $f(\alpha',\beta')=0$. Thus P' lies on C also. By the same argument $\partial f/\partial x(\alpha,\beta)=0$ if and only if $\partial f/\partial x(\alpha'\beta')=0$, and the same statement holds for $\partial f/\partial y$. Thus if P is simple, then so is P'.

(b) The proof makes heavy use of concepts from field theory and is somewhat abstract. Let (α,β) and (α',β') be conjugate points of C: $f(x,y)=0$, defined over K, and let E and E' be the fields $K[\alpha,\beta]$ and $K[\alpha',\beta']$. Every element of E can be written as $g(\alpha,\beta)$ where g is some polynomial in $K[\dot{x},\dot{y}]$. Furthermore, by definition $g(\alpha,\beta)=0$ if and only if $g(\alpha',\beta')=0$. Hence we can define a map σ from E to E' by taking $g(\alpha,\beta)$ to $g(\alpha',\beta')$. This map is an isomorphism: it is bijective and preserves addition and multiplication. Furthermore, it leaves elements of K unchanged and so the equation $f(x,y)=0$ remains unchanged.

Hence σ maps the function field $E(C)$ to $E'(C)$ leaving elements of $K(C)$ unchanged. The elements of the local ring $E[C]_P$ are those of the form $k(x,y)/h(x,y)$ with $h(\alpha,\beta)=\gamma \neq 0$ in E. These get mapped to quotients $\sigma k(x,y)/\sigma h(x,y)$ and $\sigma h(\alpha',\beta')=\sigma\gamma$ which is non-zero because σ is an isomorphism. So $E[C]_P$ gets mapped to $E'[C]_{P'}$. Therefore if $\varphi \in E(C)$ has the form $(x-\alpha)^n\psi$ with $\psi \in E[C]_P$, then $\sigma\varphi$ has the form $(x-\alpha')^n\sigma\psi$ and $\sigma\psi \in E'[C]_{P'}$. By symmetry the converse also holds and so the x-order of φ at P is the same as the x-order of $\sigma\varphi$ at P'. If $\varphi \in K(C)$, then $\varphi=\sigma\varphi$ and thus it has the same order at both points, as claimed. □

3.15 The degree of a point

To measure how much we must enlarge K to obtain the coordinates of P: (α, β), we introduce the notion of the degree of a point. It emerges that, more importantly, the degree counts the number of conjugates of P.

DEFINITION Let K be a field and P: $(x{=}\alpha, y{=}\beta)$ be a point with coefficients in a finite extension field E. Let $K[\alpha, \beta]$ be the extension field generated by the coordinates of P. Then the *degree* of P, denoted by $d(P)$, is the dimension $|K[\alpha, \beta] : K|$ of $K[\alpha, \beta]$ as a vector space over K. A point of degree 1 is called *rational.*

EXAMPLE To obtain the point P: $(\mathsf{i}, \sqrt{2})$ of the real circle $x^2 + y^2 = 1$, we must enlarge R to C. So the degree of P is 2. It has two conjugates, P itself and $\bar{P}$: $(-\mathsf{i}, \sqrt{2})$.

PROPOSITION (a) *Conjugate points have the same degree.*
(b) *If P: $(x{=}\alpha, y{=}\beta)$ is a point of degree n of the curve C defined over a finite field K, then in any extension field of $K[\alpha, \beta]$, P has exactly n conjugates and their coordinates all lie in $K[\alpha, \beta]$.*

EXAMPLE For the cubic curve $x^3 + y^3 = 1$ defined over F_2 the points $(1, 0)$, $(10, 0)$, and $(2', 5')$ given in Example 2.10 are defined in F_2, F_4, and F_8 respectively. So they have degrees 1, 2, and 3. The number of conjugates of each is the same as its degree, namely $(1, 0)$ on its own, the pair $(10, 0)$, $(11, 0)$, and the triple $(2', 5')$, $(4', 6')$, $(7', 3')$.

Remark If we drop the restriction that K is finite, then part (b) need no longer hold. In that case one can only say that P has at most n conjugates over any extension field. It is also possible to give a condition when there is an extension field in which P has the full quota of n conjugates (see Exercise 11).

Proof (a) Let P': (α', β') be a conjugate of P. In Proposition 3.14 we have seen that the map σ taking $g(\alpha, \beta)$ to $g(\alpha', \beta')$ defines an isomorphism of $K[\alpha, \beta]$ onto $K[\alpha', \beta']$ fixing K. Thus the dimensions of these two fields as vector spaces over K must be equal.

(b) I shall deal with the case that neither α nor β lies in K itself. A similar but simpler argument proves the case when one of α, β lies in K. Let γ be a primitive element of $K[\alpha, \beta]$ and, say, $\alpha = \gamma^i$, $\beta = \gamma^j$. The minimal polynomial of γ has degree n and exactly n roots $\gamma = \gamma_1, \ldots, \gamma_n$ in $K[\alpha, \beta]$. For each γ_k, the point (γ_k^i, γ_k^j) is a conjugate of P. So P has at least n conjugates in $K[\alpha, \beta]$.

Now suppose P': (α', β') is conjugate to P in some extension field E of $K[\alpha, \beta]$. The isomorphism σ of part (a) maps γ onto a primitive element γ' of $K[\alpha', \beta']$, such that $\alpha' = \gamma'^i$ and $\beta' = \gamma'^j$. But γ' has the same minimal polynomial as γ and the roots of that polynomial in E are precisely $\gamma_1, \ldots, \gamma_n$. Thus $\gamma' = \gamma_k$ for some $k = 1, \ldots, n$, and P' is one of the conjugates we calculated for $K[\alpha, \beta]$. So P has exactly n conjugates and their coordinates all lie in $K[\alpha, \beta]$. □

3.16 Exercises

1. Let D be an integral domain (so that the cancellation law holds in D). Define equality, addition, and multiplication of fractions as in Paragraph 3.4 and show that if $a/b = a'/b'$ and $c/d = c'/d'$ then $a/b + c/d = a'/b' + c'/d'$ and $(a/b)(c/d) = (a'/b')(c'/d')$. Deduce that fractions do indeed form a field with D isomorphic to the subring $a/1$.

What goes wrong if the cancellation law does not hold in D?

2. Show that if $f(\dot{x}, \dot{y})$ has degree 1, then $K[\dot{x}, \dot{y}]/f(\dot{x}, \dot{y}) \cong K[\dot{t}]$.

3. Show that if (α, β) is a singular point of the Klein quartic over some field K, then $3\beta^3 + 2\alpha = 0$ and $-2\beta^3 + \alpha = 0$. Show that unless K has characteristic 7, the only solution for these equations is $\alpha = \beta = 0$. Deduce that except for fields of characteristic 7, the Klein quartic is smooth.

4. Show that$(x{=}2, y{=}4)$ is a singular point of the Klein quartic over $\mathsf{F}_7 = \mathsf{Z}/7$.

5. Rewrite the proof of Theorem 3.12, without shifting the origin (it is necessary to replace $\dot{x}$ and $\dot{y}$ by $\dot{x} - \alpha$ and $\dot{y} - \beta$ in the appropriate places).

6. Show that the curve $x^5 - y^4 - y = 0$ is smooth except for fields of characteristic 5.

7. Show that the curve $x^7 + y^7 = 1$ is smooth except for fields of characteristic 7.

8. Let C: $f(x, y) = 0$ be an algebraic curve and let its equation in the (u, v)-system be $g(u, v) = 0$. Let P: $(x{=}\alpha, y{=}\beta)$ be a point of the curves with (u, v)-coordinates (γ, δ). Show that if P is singular with respect to $f(x, y)$ it is also singular with respect to $g(u, v)$.

9. Hermitian curves. Let K be the field F_{q^2}. Consider the two Hermitian curves C: $x^{q+1} + y^{q+1} + 1$ and D: $x^{q+1} - y^q - y$ defined in Exercise 9 of Chapter 2. Show that both these curves are smooth.

10. Complete the proof of Proposition 3.15b by proving the case where (α, β) has $\beta \in K$.

11. *For readers familiar with general field theory*
Show that a point (α, β) of degree n over $\bar{K}$ has exactly n conjugates if $K[\alpha, \beta]$ is a separable extension of K.

12. Let C: $x^2 + y^2 = 1$ be the real unit circle and let $t = y/(x + 1)$ in $\mathsf{R}(C)$. Show that $x = (1 - t^2)/(1 + t^2)$ and $y = 2t/(1 + t^2)$. Deduce that $\mathsf{R}(C)$ is isomorphic to the field of rational functions $\mathsf{R}(t)$. This isomorphism does not identify $\mathsf{R}[C]$ with the polynomial ring $\mathsf{R}[t]$.

Show that t is the slope of the line through $(-1, 0)$ and (x, y) so that t can be interpreted as $\tan(\theta/2)$, where $\tan(\theta) = y/x$.

13. Euler's substitution. Every quadratic curve is rational.

Let C: $f(x, y) = 0$ be a smooth curve with f of degree 2, such that the point P: $(0, 0)$ lies on C. Show that for any t, the equation $f(x, tx) = 0$ is a quadratic with one root $x = 0$ (except for possibly two values of t). Let u be the other root; show that there is a rational function $g(z)$ with at most two poles such that $u = g(t)$. Deduce that every (x, y)-point except P of the curve C is of the form $(g(t), tg(t))$ for some choice of t.

14. Euler's substitution (cont.). With the assumptions of Exercise 13, let $h(t) = f(g(t), tg(t))$. Show that h is the zero function in t. Now let φ be the map from $K[x,y]$ to $K(t)$ taking $k(x,y)$ to $k(g(t), tg(t))$. Show that φ induces an isomorphism of $K(C)$ with $K(t)$. This proves that C is rational.

15. Assuming that K does not have characteristic 3, show that the cubic $x^3 + y^3 - 1$ is not a rational curve.

Hint: Suppose we could embed $K[x,y]/(x^3 + y^3 - 1)$ into the rational functions in t. Then let the image of x be $p(t)/r(t)$ and y be $q(t)/r(t)$, where we can assume that the polynomials $p(t), q(t), r(t)$ have no common factors. Then in $K[t]$

$$p^3(t) + q^3(t) - r^3(t) = 0.$$

Differentiate this equation, cancel the 3, and get

$$\begin{bmatrix} p(t) & q(t) & r(t) \\ p'(t) & q'(t) & r'(t) \end{bmatrix} \begin{bmatrix} p^2(t) \\ q^2(t) \\ r^2(t) \end{bmatrix} = \begin{bmatrix} 0 \\ 0 \end{bmatrix}.$$

Deduce first that p^2, q^2, and $-r^2$ are rational multiples of $qr' - q'r$, $rp' - r'p$, and $pq' - q'p$ respectively, and then since p, q, and r are relatively prime, that $p^2 \mid qr' - q'r$, $q^2 \mid rp' - r'p$, and $-r^2 \mid pq' - p'q$. Let, say, p have the largest degree among p, q, and r; then $2\deg(p) \le \deg(q) + \deg(r) - 1$, which is a contradiction.

4

A SURVEY OF THE THEORY OF ALGEBRAIC CURVES

In this chapter I shall state the main theorems of the theory of algebraic curves up to Riemann's theorem. These theorems are important in constructing and establishing the properties of geometric Goppa codes, but they do not enter directly into the calculations, once the codes have been constructed. As the proofs of the principal theorems are deep, I defer them to Part II of the book. Here I shall show by means of examples and consequences derived from the results how they enable us to analyse the properties of curves.

4.1 Existence of zeros

We begin with two fundamental theorems, which are generalizations of important theorems of complex number theory. These state that all non-constant functions have zeros and poles, and furthermore they have the same number if they are counted correctly.

THEOREM Zeros theorem *Let C: $f(x,y)=0$ be an affine curve over K and let $g(\dot{x},\dot{y})$ be a polynomial in $K[\dot{x},\dot{y}]$. If $g(\dot{x},\dot{y})$ is not congruent to a constant modulo $f(\dot{x},\dot{y})$, then there exists a point P: $(x{=}\alpha, y{=}\beta)$ of C with α,β in some finite extension field of K, such that $g(\alpha,\beta)=0$. If $g(\dot{x},\dot{y})$ is not congruent to 0 modulo $f(\dot{x},\dot{y})$, then the number of such points is finite.* □

The zeros theorem has a number of important corollaries.

COROLLARIES (a) *If $g(\dot{x},\dot{y})$ is not congruent to a constant modulo $f(\dot{x},\dot{y})$ then for any value γ in a finite extension field E of K there exists a point P: $(x{=}\alpha, y{=}\beta)$ of C with α,β in some finite extension field of E such that $g(\alpha,\beta)=\gamma$.*
(b) *C has infinitely many points.*
(c) *If $g(\dot{x},\dot{y})$ and $f(\dot{x},\dot{y})$ are absolutely irreducible and $g(\dot{x},\dot{y})$ is not a constant multiple of $f(\dot{x},\dot{y})$, then the curves C: $f(x,y)=0$ and D: $g(x,y)=0$ are distinct.*
(d) *C has only finitely many singular points.*

Proofs (a) Consider C as a curve defined over E. Then the polynomial $g(\dot{x},y)-\gamma$ lies in $E[\dot{x},\dot{y}]$ and is not congruent to a constant modulo $f(\dot{x},\dot{y})$ (see Exercise 9). Hence for some point P: $(x{=}\alpha, y{=}\beta)$ of C, we have $g(\alpha,\beta)-\gamma=0$.

(b) There are infinitely many values γ in extension fields of K and so we need only show that there exists a polynomial not congruent to a constant modulo $f(\dot{x},\dot{y})$. If the degree of $f(\dot{x},\dot{y})$ is greater than 1 we can take $g(\dot{x},\dot{y})=\dot{x}$. Otherwise $f(\dot{x},\dot{y})=a\dot{x}+b\dot{y}+c$ and we can take $g(\dot{x},\dot{y})=\dot{x}^2$.

(c) If the curves are the same, then $g(\dot{x}, \dot{y})$ has infinitely many zeros on the curve defined by $f(\dot{x}, \dot{y})$. From the theorem it follows that $g(\dot{x}, \dot{y})$ is a multiple of $f(\dot{x}, \dot{y})$. By symmetry $f(\dot{x}, \dot{y})$ is also a multiple of $g(\dot{x}, \dot{y})$. Hence they differ by a constant factor.

(d) The singular points of C are the common zeros of the two partial derivatives on C. Since the partial derivatives have degree less than $\deg f(\dot{x}, \dot{y})$ each can only be congruent to a constant if it is equal to that constant. If either partial derivative is non-constant then the theorem implies that it has only finitely many zeros. If it is a non-zero constant, then it has no zeros. Thus there can only be infinitely many singular points if both partial derivatives are zero. That can only occur if all the terms of f have the form $ax^{pi}y^{pj}$, where p is the characteristic of K. In that case we can extend K to a field E which contains the pth roots of all the coefficients of f. Over E we can write $f = g^p$, where g has a term $\sqrt[p]{a}x^i y^j$ for each term $ax^{pi}y^{pj}$ of f. That contradicts the absolute irreducibility of f.

□

4.2 Counting poles and zeros

For most purposes conjugate points should not be distinguished. So we coagulate conjugate points. As we can only define orders for non-singular points, we restrict our attention to these.

DEFINITION Let C be a curve and P a non-singular point of C; the set of conjugates of P is called a *point class*. We abuse notation and denote the point class of P by P also. As the degrees and order functions of conjugate points are the same, we can speak of the *degree* or *order* function of a point class, and use the notation $d(P)$ and $\nu_P(\theta)$ equally for point classes or their member points.

With this concept we can formulate a grand extension of the theorem that the number of roots of a polynomial of degree n, counted with the correct multiplicity, is exactly n. For this theorem we assume that the curve we are dealing with is smooth, that is that all its points are non-singular.

THEOREM Degree theorem *Let C: $f(x, y) = 0$ be a smooth curve defined over K, and let $\varphi \in K(C)$. Then $\nu_P(\varphi) \neq 0$ for only finitely many point classes P, and summing over all point classes we have*

$$\sum \nu_P(\varphi) d(P) = 0.$$

□

COROLLARY *Replacing each point class P by its $d(P)$ constituent points we see that if we sum over all non-singular points the sum becomes*

$$\sum \nu_P(\varphi) = 0.$$

That states that if the poles and zeros of φ are counted with the correct order, then φ has equally many poles and zeros.

EXAMPLE It is interesting to confirm what the theorem says for ordinary polynomials. Consider the curve $y = 0$. This straight line has as its function field ordinary rational functions in one indeterminate. It has points $(\alpha, 0)$ with α in some extension field and there is a single point at infinity $(u{=}0, v{=}0)$. It has degree 1 (because its coordinates lie in the base field) and its order function assigns to each polynomial $f(x) = f(1/u)$ the value $-\deg(f)$.

Two finite points $(\alpha, 0)$ and $(\beta, 0)$ are conjugate if α and β have the same minimal polynomial. Thus finite point classes correspond to irreducible polynomials in $K[\dot{x}]$ and the degree of the point class P corresponding to the irreducible $p(\dot{x})$ is just $\deg(p)$. The order of a polynomial $f(\dot{x})$ at this point class is the power to which $p(\dot{x})$ occurs in the (unique) factorization of $f(\dot{x})$.

As there are only finitely many irreducible factors of f the number of point classes where f has order > 0 is finite. Furthermore the sum

$$\sum \nu_P(f) d(P)$$

over these point classes is precisely the degree of f. If we add in the point at infinity, then $\nu_\infty(f) = -\deg(f)$ and so the sum becomes 0.

4.3 The horizon theorem

In defining a code based on a curve C we shall need to use linear subspaces of the field of functions $K(C)$. The most natural way to specify such a space is to require its members to have specified orders at certain point classes.

EXAMPLE For the field $K(x)$, the set of rational functions with order $\geq -n$ at ∞ is the vector space of rational functions of degree $\leq n$. This is infinite dimensional.

If we also require the function f/g to have no poles at any finite point classes, that means that no irreducible polynomial divides g. Thus g is a constant and we now have the set of polynomials of degree $\leq n$. This has dimension $n + 1$.

Suppose we relax the restriction and allow order ≥ -1 at the point $x = 0$. Now we have the set of functions of the form f/x where f has degree $\leq n + 1$. The dimension increases to $n+2$. If we tighten the condition instead and require a zero at $x = 0$, we get the set of polynomials of the form xf with f of degree $< n$. The dimension drops to n. Of course, if we require the order to be at least 0 at all finite point classes and > 0 at ∞ we are asking for a function f/g with $\deg(g) > \deg(f)$ and no poles. There are no such functions (except 0, which we give infinite order at all point classes) and the dimension is 0.

To perform the corresponding calculations over an arbitrary curve we need a further theorem, which gives us a criterion for a function in $K(C)$ to be representable by a polynomial.

THEOREM Horizon theorem *Let C: $f(x, y) = 0$ be an (x, y)-smooth curve and let φ be a rational function with $\nu_P(\varphi) \geq 0$ for all (x, y)-points P: $(x{=}\alpha, y{=}\beta)$ of C. Then in $K(C)$, $\varphi \equiv p(x, y)$ for some polynomial $p(\dot{x}, \dot{y}) \in K[\dot{x}, \dot{y}]$.*

Informally, a function with poles only on the horizon is a polynomial. □

COROLLARY *A function in $K(C)$ without any poles is a constant. A function without any zeros is a constant.*

Proof By the degree theorem a function has no poles if and only if it has no zeros. If g/h has no poles, then by the horizon theorem, g/h is congruent to a polynomial $p(\dot{x}, \dot{y})$ modulo $f(\dot{x}, \dot{y})$. Since g/h has no zeros either $p(x, y)$ has no zeros. Then by the zeros theorem, $p(\dot{x}, \dot{y})$ is congruent to a constant modulo $f(\dot{x}, \dot{y})$.

4.4 Applying the horizon theorem

The horizon theorem is very useful for determining the nature of functions with specified poles and zeros.

EXAMPLE Consider the curve $C\colon x^3 + y^3 = 1$ defined over F_2 and the point $P\colon (x{=}0, y{=}1)$. What is the dimension of the set of rational functions $f/g \in K(C)$ with $\nu_P(f/g) \geq -n$ and $\nu_Q(f/g) \geq 0$ everywhere else?

If $n = 0$, then the space consists of the constants and the dimension is 1.

If $n = 1$, then the space still consists of constants.

You can see this as follows. A function satisfying our condition must be of the form f/g with $g(0,1) \neq 0$. Now since $\nu_P(x) = 1$, and $\nu_Q(x) \geq 0$ for all points $Q\colon (x{=}\alpha, y{=}\beta)$, xf/g satisfies the conditions of the horizon theorem. Thus xf/g is congruent to a polynomial $h(x, y)$ in $K(C)$ and our function takes the form $h(x, y)/x$. Let $\deg(h(x, y)) = d$. In the (u, v)-system the function transforms to $h(1/v, u/v)v$, which involves a term au^j/v^{d-1}. If $d > 1$ it therefore has a prohibited pole at $(u{=}1, v{=}0)$. So $\deg(h(\dot{x}, \dot{y})) \leq 1$ and our function must be of the form $(ax + by + c)/x$.

To prevent a pole at $(x{=}0, y{=}10)$, we must have $10b + c = 0$, but to prevent a pole at $(x{=}0, y{=}11)$, we must have $11b + c = 0$. Hence $b = c = 0$ and our function is the constant $a = ax/x$.

Thus there are no non-constant functions of this type and the dimension remains 1.

If $n = 2$ we have a function $x/(y + 1) = y^2 + y + 1/x^2$ which has a pole of order 2 at P. The function has zeros at $(x{=}0, y{=}10)$ and $(x{=}0, y{=}11)$. At all other points the order of this function is 0. So the dimension of our space for $n = 2$ is at least 2.

If $n = 3$ there is a further function $1/(y + 1) = y^2 + y + 1/x^3$. This has a pole of order 3 at P. It has no further zeros or poles in the (x, y)-system. In the (u, v)-system it transforms to $v^2u + vu^2 + u^3$ which has zeros at $(u{=}0, v{=}1)$, $(u{=}0, v{=}10)$, and $(u{=}0, v{=}11)$. The dimension of our space for $n = 3$ is at least 3.

We shall shortly find that the dimensions of these spaces for $n = 2, 3$ are exactly 2 and 3.

4.5 Divisors

We can combine conditions at several points into a single condition by defining a divisor. We use point classes to ensure that conjugate points are treated equally.

DEFINITION A *divisor* D of a curve C assigns an integer value $D(P)$ to every point class of C. We require that $D(P) \neq 0$ for only finitely many point classes.

We shall use the notation $D = \sum D(P)P$ to denote the divisor. Addition of divisors is defined term by term, and we say $D \leq E$ if $D(P) \leq E(P)$ for all P. If only one point class P has a possibly non-zero value $D(P) = n$, we write $D = nP$.

Let S be a set of point classes and let D be a divisor. The *L-space* $L(D, S)$ is the set of elements $\varphi \in K(C)$ such that $\nu_P(\varphi) + D(P) \geq 0$ for all $P \in S$. If S is the set of all point classes of C it is omitted.

EXAMPLE In Example 4.3 we first considered the straight line $y = 0$. We calculated $L(nP, S)$ where $P = \infty$ and $S = \{P\}$. It is the space of all rational functions of degree $\leq n$. On the other hand, we found that $L(nP)$ is the space of polynomials of degree $\leq n$.

In Example 4.4, we then considered the curve $x^3 + y^3 = 1$. For P: $(0, 1)$ we found that $L(0P) = L(P)$ and that both were just the set of constants. We found non-trivial elements of $L(2P)$ and $L(3P)$.

THEOREM *The L-space $L(D, S)$ is a vector space over K. If S is finite, then $\dim(L(D, S)) = \infty$, but $\dim(L(D))$ is finite.* □

4.6 A special case

Although we cannot prove Theorem 4.5 here, we shall prove a special case that will suffice for our calculations.

PROPOSITION *Let C be a smooth curve defined over K, and let P be a point of C with coefficients in K. Let D be a divisor of the form nP with $n \geq 0$. Suppose $1 = \varphi_1, \varphi_2, \ldots, \varphi_k \in L(D)$ with $\nu_P(\varphi_j) = -n_j$. Suppose that*

(i) $0 = n_1 < n_2 < \cdots < n_k \leq n$,

(ii) *If $\psi \in L(D)$ then $\nu_P(\psi) = n_j$ for some $j = 1, \ldots, k$.*

Then $\varphi_1, \ldots, \varphi_k$ forms a basis of $L(nP)$.

EXAMPLE Using the proposition, we can now complete the calculations of Example 4.4. Let C: $x^3 + y^3 = 1$ over F_2 and P: $(0, 1)$. We have shown that there are no functions ψ with $\nu_P(\psi) = -1$ and $\nu_Q(\psi) \geq 0$ for $Q \neq P$. But we have given examples of functions with $\nu_P(\psi) \leq -2$ and $\nu_Q(\psi) \geq 0$ elsewhere. Indeed, if we take the functions $\varphi = x^i y^j / (y + 1)^{i+j}$, then as $\nu_P(x) = 1$, $\nu_P(y) = 0$, and $\nu_P(y + 1) = 3$, we have $\nu_P(\varphi) = -2i - 3j$. Choosing appropriate non-negative values i and j will give every value strictly less than -1.

We must check that the functions have no further poles. As $(0, 1)$ is the only point in the (x, y)-system with $y = 1$, φ has no further poles in that system. In the (u, v)-system φ has the form $u^j / (v + u)^{i+j}$. As the curve has the (u, v)-equation $u^3 + v^3 = 1$, there are no poles in the (u, v)-system. The only point of the (w, z)-system we need to consider is $(0, 0)$, but this does not lie on C. That proves that the only pole of φ is P, and thus $\varphi \in L((2i + 3j)P)$.

Applying the proposition to functions φ with $\nu_P(\varphi) = 0, 2, \ldots, n$, we find that $L(nP)$ has dimension n for $n \geq 1$, but $L(0P)$ has dimension 1.

Proof For convenience we assume that P is the (x, y)-point $(x=\alpha, y=\beta)$ with $\alpha, \beta \in K$. As you are asked to show in Exercise 5, the order of a linear combination of $\varphi_1, \ldots, \varphi_j$ at P is at least $-n_j$ while φ_{j+1} has order $< -n_j$. So φ_{j+1} is not a linear combination of $\varphi_1, \ldots, \varphi_j$. Hence the functions φ_i are linearly independent.

To prove that they form a basis, we show that every function $\psi \in L(nP)$ is a linear combination of $\varphi_1, \ldots, \varphi_k$ by the induction on $-\nu_P(\psi)$. If $\nu_P(\psi) \geq 0$, then since $\nu_Q(\psi) \geq 0$ everywhere else, ψ is a constant and so $\psi = a\varphi_1$. Suppose the proposition is true for $\nu_P(\psi) > -n_j$ and suppose we have $\psi \in L(D)$ with $\nu_P(\psi) = -n_j$. Then $\nu_P(\psi/\varphi_j) = 0$. So the value of ψ/φ_j at P is a non-zero constant a in K. Let $\chi = \psi/\varphi_j - a$. Then $\nu_P(\chi) > 0$, and so $\nu_P(\chi\varphi_j) > -n_j$. But $\chi\varphi_j = \psi - a\varphi_j$, and since ψ and $\varphi_j \in L(nP)$ it follows that $\chi\varphi_j \in L(nP)$. By induction hypothesis,

$$\chi\varphi_j = a_1\varphi_1 + \cdots + a_{j-1}\varphi_{j-1}.$$

So

$$\psi = a_1\varphi_1 + \cdots + a_{j-1}\varphi_{j-1} + a\varphi_j. \qquad \square$$

4.7 Riemann's theorem and the genus

Riemann (1826–1866) developed the idea of the genus of a curve in the 1850s. His concept of the genus was geometric rather than algebraic and based on the structure of the real surface defined by a complex curve. He considered such a surface to be a covering surface of the complex plane or its compactification the Riemann sphere. The genus then describes the geometric nature of this Riemann surface. The genus forms a far more powerful classification tool for surfaces than the degree. Riemann himself found algebraic methods for calculating the genus and these can be adapted to curves over other fields, giving a purely algebraic definition of the genus. That is the method adopted here.

DEFINITION For a divisor $\sum D(P)P$, the dimension of $L(D)$ is called the *rank* of D and denoted by $l(D)$. The value $\sum D(P)d(P)$, where $d(P)$ is the degree of the point class P, is called the *degree* of D and denoted by $d(D)$.

Calculating the value of $l(D)$ is quite difficult, as you have seen. It does not behave as regularly as one might expect. For instance, we found that $l(P) = 1 = l(0P)$ for $x^3 + y^3 = 1$ and $P = (0, 1)$. In general, the behaviour of the rank becomes regular as the degree $d(D)$ becomes large. Riemann's theorem, which is the most important result of the theory of algebraic curves, gives a general estimate $l(D)$ in terms of a constant γ, called the genus of the curve. Using this theorem greatly simplifies the calculations required to determine $L(D)$.

THEOREM Riemann's theorem *Let C be an algebraic curve over K.*
(a) *Let A and B be divisors on C with $A \leq B$ (that is. $A(P) \leq B(P)$ for all points P of C). Then $l(B) - d(B) \leq l(A) - d(A)$.*
(b) *There exists a non-negative number γ such that $l(D) - d(D) = 1 - \gamma$ for all divisors D with $d(D) > 2\gamma - 2$.* $\square$

The algebraic proof of Riemann's theorem only shows that the equation $l(D) - d(D) = 1 - \gamma$ holds for large enough degree. The precise bound is a consequence of the Riemann–Roch extension proved by Riemann's pupil Roch.

In the special case that $D = nP$, Proposition 4.6 provides a partial proof of the first part of Riemann's theorem.

PROPOSITION *Let C be a smooth curve defined over K and let P be a point with coefficients in K. Then $l(nP) \leq l((n+1)P) \leq l(nP) + 1$.*

Proof Choose a basis of $L((n+1)P)$ satisfying the conditions of Proposition 4.6. Then it contains at most one function with $\nu_P(\varphi) = n + 1$. Removing this function (if it is present) produces a basis of $L(nP)$. □

DEFINITION The number γ found in Riemann's theorem is called the *genus* of the curve C.

The following two corollaries are immediate consequences of the theorem.

COROLLARIES (a) *The genus of C is unique.*
(b) *For all divisors D on C, $l(D) - d(D) \geq 1 - \gamma$.*

EXAMPLE Example 4.4 corroborates the fact that the straight line $y = 0$ has genus 0. Any quadratic curve has a function field isomorphic to ordinary rational functions. It follows that quadratic curves also all have genus 0.

From the fact that $l(P) = 1 = d(P)$ for the point P: $(0,1)$ of the curve $x^3 + y^3 = 1$ over F_2, it follows that this curve cannot have genus 0. The calculations in Example 4.6 show that this curve has genus 1.

4.8 The Plücker formula for smooth plane curves

The examples we have calculated hint at the idea that the genus of a cubic curve ought to be 1, and one might guess that maybe there is a formula for the genus in terms of the degree of the defining equation. That is indeed the case for smooth curves. The formula is known as the Plücker formula, but what Plücker himself proved was a set of equations relating the degree of a plane curve to its singularities of different types. He published these equations in 1834 about 20 years before Riemann introduced the idea of the genus. What is now known as the Plücker formula is a simplification of Plücker's equations using the genus. We shall only state the formula for the case when there are no singularities. You can find a fuller version in Fulton (1989).

THEOREM Plücker formula *Let C: $f(x, y) = 0$ be a smooth curve of degree n. Then the genus of C is $(n-1)(n-2)/2$.* □

As expected, the formula gives the genus of curves of degree 1 and 2 as 0, and the genus of a cubic curve as 1.

In the next paragraphs we shall use the theorem to perform calculations for the Klein quartic and the curve $x^5 + y^5 = 1$ similar to those we performed for $x^3 + y^3 = 1$. For completeness we first tabulate the results of Example 4.6.

EXAMPLE Let C be the curve $x^3 + y^3 = 1$ and $P = (0, 1)$. The genus of C is 1 and the values $l(nP)$ are as follows:

n	0	$n > 0$
$l(nP)$	1	n

4.9 The Klein quartic

EXAMPLE The Klein quartic over F_4.

The equations in the three systems are $x^3y + y^3 + x = 0$, $v^3u + u^3 + v = 0$, $w^3z + z^3 + w = 0$. By the Plücker formula the curve has genus 3.

Choose as our base point P the point $(x{=}0, y{=}0)$. We have $\partial f/\partial y(0,0) = 0$. So we must use the y-order. Thus $\nu_P(y) = 1$ and as $x(x^2y+1) = y^3$, $\nu_P(x) = 3$.

Consider the function $\varphi = y^i/x^j$ with $0 \leq 3i \leq 2j$. This function has a pole of order $3j - i$ at P. Other (x, y)-points on C have $x \neq 0$ and are not poles, but we must also consider (u, v) points with $v = 0$ and the (w, z) point Q: $(w{=}0, z{=}0)$. In the (u, v)-system y^i/x^j becomes $v^{j-i}u^i$ which has no pole because $j \geq i$. In the (w, z) system it becomes w^{j-i}/z^j. Since $\nu_Q(w) = 3$ and $\nu_Q(z) = 1$, $\nu_Q(w^{j-i}/zj) = 2j - 3i \geq 0$. So φ has no further poles.

The values $\nu_P(y^i/x^j) = 3j - i$, we can obtain for these functions are 0, 3, 5, and all numbers > 5. So for $n \geq 5$ we get $n - 2$ values. From Riemann's theorem it follows that for $n \geq 5$ the functions y^i/x^j with $0 \leq 3i \leq 2j$ form a basis of $L(nP)$.

Suppose $n < 5$. If ψ has a pole only at P then ψ certainly also lies in $L(5P)$. Thus $\nu_P(\psi) = \nu_P(y^i/x^j)$ for some admissible values i, j. By Proposition 4.6 it follows that appropriate functions $\varphi = y^i/x^j$ still form a basis of $L(nP)$. We can now write out a table of values for $l(nP)$.

n	0	1	2	3	$n \geq 4$
$l(nP)$	1	1	1	2	$n-2$

4.10 A Hermitian quintic

EXAMPLE The curve $x^5 + y^5 = 1$ over F_{16}.

This curve has genus 6.

We take as our base point P the point $(x{=}0, y{=}1)$. As $\partial f/\partial x(0,1) = 0$, we use the x-order. Thus $\nu_P(x) = 1$ and $(y+1)(y^4 + y^3 + y^2 + y + 1) = x^5$, so $\nu_P(y+1) = 5$.

The function $\varphi = x^iy^j/(y+1)^{i+j}$ has $\nu_P(\varphi) = -(4i + 5j)$. Just as in the case of $x^3 + y^3 = 1$ we note that $x^iy^j/(y+1)^{i+j} = u^j/(u+v)^{i+j}$ which has no poles with $v = 0$ and $u^5 = 1$. Since $(w{=}0, z{=}0)$ is not a point of C it follows that this function has no further poles. The values of n representable as $4i + 5j$ with $i, j \geq 0$ are 0, 4, 5, 8, 9, 10, and all values ≥ 12. For $n > 10$ the number of such values is $n - 5$.

By Riemann's theorem it follows that appropriate functions $\varphi = x^iy^j/(y+1)^{i+j}$ with $\nu_P(\varphi) \geq -n$ form a basis of $L(nP)$ for $n > 10$. It follows that if $\psi \in K(C)$ is a function with a pole only at P, then $\nu_P(\psi) = 4i + 5j$ for some

i and j. Hence by Proposition 4.6, appropriate functions $\varphi = x^i y^j/(y+1)^{i+j}$ form a basis of $L(nP)$ for all $n \geq 0$. The table below lists $l(nP)$.

n	0	1	2	3	4	5	6	7	8	9	10	$n > 10$
$l(nP)$	1	1	1	1	2	3	3	3	4	5	6	$n-5$

Just as we proved the dimension formulae for $x^3 + y^3 = 1$ directly, it is possible to do so for Examples 4.9 and 4.10 also, but it is much easier to apply Riemann's theorem.

In the next chapter we shall use the spaces $L(D)$ to construct Goppa codes.

4.11 Exercises

1. Why do the coordinates of conjugate points of an algebraic curve have the same minimal polynomials?
2. Let K be a field and α, β be algebraic elements of an extension E. Let the degrees of the minimum polynomials of α and β be m and n respectively. Show that the elements $\alpha^i\beta^j$, for $0 \leq i \leq m$ and $0 \leq j \leq n$, form a basis of $K[\alpha, \beta]$.
3. Prove that if P: (α, β) is a point of degree n over a finite field K, with $\alpha \in K$, then P has exactly n conjugates in any extension field of $K[\beta]$.
4. Show that the conjugates of the point (α, β) defined over a field of characteristic 2 are (α, β), (α^2, β^2), (α^4, β^4), $\ldots$.
5. Verify the claim in Proposition 4.6 that the order of a linear combination of functions φ_i at a point is at least equal to the minimum of the orders of the individual functions φ_i.
6. Show that for any curve C and any point P of degree 1 on C, the rank $l(nP)$ is at most $n+1$. Deduce that no curve has negative genus.
7. Let C be a curve with a point P of degree 1 such that $l(P) = 2$. Show that $l(nP) = n+1$ for all n and deduce that the genus of C is 0.
8. Let P be the point $(x{=}0, y{=}0)$ on the Klein quartic $x^3y + y^3 + x = 0$, and let Q be the point $(x{=}0, y{=}1)$ on the quintic $x^5 + y^5 + 1$. Show directly that functions in $L(P)$ and $L(Q)$ are constants. Deduce this also from Exercise 7.
9. Let $f(\dot{x}, \dot{y})$, $g(\dot{x}, \dot{y})$ be non-constant polynomials in $K[\dot{x}, \dot{y}]$, and let E be an extension field of K. Suppose that $g(\dot{x}, \dot{y}) - q(\dot{x}, \dot{y})f(\dot{x}, \dot{y})$ is a constant α for some polynomial $q(\dot{x}, \dot{y}) \in E[\dot{x}, \dot{y}]$. Show that $q(\dot{x}, \dot{y}) \in K[\dot{x}, \dot{y}]$ and hence $\alpha \in K$.

   Hint: Suppose that $q(\dot{x}, \dot{y}) \notin K[\dot{x}, \dot{y}]$. Let k be the highest degree of a term in $q(\dot{x}, \dot{y})$ with coefficient not in K and let $d = \deg f(\dot{x}, \dot{y})$. Show that the term of degree $k+d$ in $g(\dot{x}, \dot{y}) - q(\dot{x}, \dot{y})f(\dot{x}, \dot{y})$ does not lie in K, and hence cannot be 0.
10. Let C be the x-axis defined by $y = 0$ over F_{16} and denote the point $(x{=}\beta, y{=}0)$ by (β) and the point $(u{=}0, v{=}0)$ by (∞) and let $D = a(0) - b(\infty)$. Show that the functions x^i with $-a \leq i \leq -b$ form a basis of $L(D)$.
11. Let C be as in Exercise 10, and let $g(x)$ be an irreducible polynomial of degree d. Let P be the point class of C associated with $g(x)$ (see Example 4.2). Show that if $D = aP - b(\infty)$, then the functions $x^i g^j$, with $i \geq 0$, $j \geq -a$, and $dj + i \leq -b$, form a basis of $L(D)$.

12. Find the genus for the curves $x^5 + y^4 + y = 0$ and $x^7 + y^7 = 1$ over F_2.
13. Verify that $x^i y^j/(y+1)^{i+j}$ transforms to $u^j/(u+v)^{i+j}$ in the (u, v)-system.
14. Let P be the point $(x{=}0, y{=}1)$ of $x^7 + y^7 = 1$. Calculate the dimensions $l(nP)$ for all n.
15. Let Q be the unique point at infinity of the curve $x^5 + y^4 + y = 0$. Show that $\nu_Q(x) = -4$ and $\nu_Q(y) = -5$ and that Q is the only pole of these functions. Use this fact to construct a basis of $L(nQ)$ consisting of functions of the form $x^i y^j$. Show that the condition $0 \leq i \leq 4$ uniquely determines one such basis.

5

GEOMETRIC GOPPA CODES

Goppa's discovery of his codes based on algebraic curves has prompted a great volume of work both in coding theory and algebraic geometry (see Goppa 1981a, 1981b). The reason is that with these codes it is possible to describe sequences of codes which are dramatically better than any previously constructed codes. That topic is expanded in Chapter 15. This chapter discusses the construction of the Goppa codes, and the next two describe error processing techniques for the codes. We begin with a short recapitulation of the basic concepts for error-correcting codes. For a more detailed introduction the reader is referred to my book (Pretzel 1996). For the purposes of this book we restrict our attention to *linear block* codes.

5.1 Error-correcting codes

DEFINITION A *code* C over the finite field K is a subspace of K^n. Its *block length* is n, its *rank* is its dimension. The vectors of K^n are called *words* or *code words* if they lie in C.

The model behind these concepts is the following. A message consisting of elements of K is to be transmitted via some medium (called the *channel*). It is first divided into message words of length equal to the rank of the code. Before it is transmitted, each message word is transformed into a code word by an invertible linear transformation. If a code word is received without distortion it can be transformed back into the corresponding message word. However, the additional information introduced by the transformation into a code word will allow the message to survive a certain amount of distortion.

DEFINITION For a code C of rank m and block length n an $n \times m$ matrix G is called a *generator matrix* if the transformation from K^m to K^n defined by $u \mapsto Gu$ has C as its image. An $(n-m) \times n$ matrix H is called a *check matrix* for C if $Hu = 0$ holds if and only if u is a code word.

If a generator or check matrix contains a submatrix of full rank that is an identity matrix, then it is said to be in *standard form.*

With a generator matrix for a code it is possible to perform the encoding and decoding that converts a message word to a code word and back. For matrices in standard form the decoding step is particularly simple. It consists of extracting the entries of a code word corresponding to the columns of the identity matrix. Otherwise decoding v involves solving the linear equations $Gu = v$. Simple decoding is desirable, because the receiver will also need to check for errors and if possible correct them. The check matrix provides an easy test for errors.

Since either a generator matrix or a check matrix determines the code, it is not surprising that one can construct a check matrix from a generator matrix and vice versa. It is particularly easy to do this for standard form matrices.

PROPOSITION *Let C be a code of rank m and block length n with generator matrix G in standard form. Assume that permuting the rows by a permutation σ brings G into the form $\binom{I}{A}$ where I is an $m \times m$ identity matrix. If J is an $(n-m) \times (n-m)$ identity matrix, then permuting the columns of $(-A, J)$ by σ^{-1} produces a check matrix H for C.*

Proof This is simple linear algebra. Let P be the $n \times n$ matrix obtained from the identity matrix by permuting its rows by σ. Then $\binom{I}{A} = PG$ and $(-A, J) = HP^{\top}$. The matrices P and $P^{\top}$ are inverses. So $HG = -AI + JA = 0$ and thus $Hv = 0$ for all code words. Since H contains an identity matrix it has rank $n - m$ and hence by the rank and nullity theorem its null space has dimension m. As C is of dimension m and lies in the null space, it must be the full null space. □

5.2 Weight and distance

The check matrix of a code only tests whether a word is a code word or not. It does not perform any form of correction. To do that we must decide what the most likely transmitted word for any given received word is. It is natural to choose the word which requires the least correction. After all, if you read the word 'theroem' it is more likely that I intended 'theorem' than 'proposition'. That leads to the following definition.

DEFINITION The *weight* $\mathrm{wt}(u)$ of a word u is the number of non-zero entries in u. The (Hamming) *distance* $d(u, v)$ between two words is the number of entries in which they differ, or $\mathrm{wt}(u - v)$. The *minimum distance* of the code C is the smallest distance between two distinct code words. This is also the smallest weight of a non-zero code word.

Transmission distortion is modelled by assuming that an error word e is added to the code word c to produce the received word v. The weight of e is called the weight of the error. If the distortion converts a code word into a non-code word it is said to be *detected*, since the check matrix will signal an error. If there is a unique code word at distance $\leq t$ from the received word, then it is possible to *correct* the error by assuming that was the transmitted code word. The fundamental fact about codes is now the following simple proposition.

PROPOSITION *Let C be a code with minimum distance d. Then it is possible to detect all errors of weight less than d. It is also possible to correct all errors of weight less than $d/2$.*

Remark It is *not* possible to do both at once, because an error pattern of weight $d - 1$ may convert the code word u to a word within distance 1 of the code word v. Then correction will take the result to v and not detect the error accurately.

Proof By the definition of minimum distance a non-zero error of weight less than d cannot convert a code word into another code word.

There cannot be two distinct code words v_1, v_2 of distance less than $d/2$ from a word u. For then $d(v_1, v_2) < d$, contradicting the assumption that d is the minimum distance of the code. □

Thus the first fundamental problem in the design of error-correcting codes is to find codes with a large rank and minimum distance in relation to the block length. In Chapter 15 a lower bound for the theoretically achievable rank and minimum distance (the Gilbert bound) will be discussed.

If a code of minimum distance $d = 2t + 1$ is used to transmit a message it is possible to correct all errors of weight at most t by using search methods, because there are only finitely many words of length n. However, that approach is not practical and it is necessary to find some algorithmic correction scheme before a code can really be used. So the second basic problem in the design of error-correcting codes is to devise efficient error processing algorithms.

5.3 Dual Goppa codes

Goppa actually introduced two dual classes of codes. It turns out that these classes are different representations of the same codes, but that is not obvious from their description, and error processors do depend on the representation of the code that is used. For the time being we shall treat the two classes as distinct. You can find a proof that they represent the same codes in Chapter 12.

I shall begin by defining 'dual' Goppa codes, and then use these to define the 'primary' Goppa codes that are used by most error processors. The reader familiar with standard coding theory will see that geometric Goppa codes are a generalization of classical Goppa codes. Together with the definitions I shall give estimates for the rank and minimum distance of the codes. These estimates are called the *designed* rank and minimum distance of the codes. Note, however, that in order to use a code, its rank must be known precisely. On the other hand a good lower bound for the minimum distance is acceptable.

DEFINITION Let K be a finite field and let C be a smooth algebraic curve defined over K. Let $\{P_1, \dots, P_n\}$ be a set of rational points (that is, points P_j of C of the form (α_j, β_j) with $\alpha_j, \beta_j \in K$), and let B be the divisor that is the sum of these places, $\sum P_j$. Further let D be a divisor such that $D(P_j) = 0$ for all $j = 1, \dots, n$. The *dual Goppa code* $C_L(B, D)$ is defined as the set of vectors $(d_1, \dots, d_n)$ such that there exists a rational function $\varphi \in L(D)$ with $d_j = \varphi(P_j)$. Such a code is usually called a Goppa *function* code.

We need to choose the points P_j to be of degree 1, so that the values d_j lie in the field K. Points of degree 1 form conjugacy classes on their own, so the divisor B is correct as it stands. Since B determines the set of points P_i we shall write $P_i \in B$ to indicate that P_i is one of the selected points.

EXAMPLE To avoid confusion, I shall use only the field F_{16} and its subfield F_4 in the examples of this chapter. You can tell if we are working over F_4 by the fact that the values are restricted to 0, 1, 10, and 11.

Table 5.1 Points of $x^3 + y^3 + 1 = 0$ over F_{16}

0: $(x=0, y=1)$, 1: $(x=0, y=10)$, 2: $(x=0, y=11)$,
3: $(x=1, y=0)$, 4: $(x=10, y=0)$, 5: $(x=11, y=0)$,
6: $(u=1, v=0)$, 7: $(u=10, v=0)$, 8: $(u=11, v=0)$.

Consider the curve C: $x^3 + y^3 = 1$ defined over $K = \mathsf{F}_4$. In Paragraph 2.10 we found all the points of C over F_4. They are the same as the points over F_{16}. There are nine points, which we number as in Table 5.1

We take as B the sum of the points 1 to 8 and as D a multiple aP_0. In Example 4.6 it was shown that the functions $x^i y^j/(y+1)^{i+j}$ have poles of order $-(2i+3j)$ at P_0 and belong to $L(D)$ if $2i + 3j \leq a$. It was also shown that choosing one such function of each order gives a basis of $L(D)$.

Table 5.2 shows the values at P_i for a choice of such functions of orders down to -7. Notice that there is no function of order -1. Transposing the entries in the rows up to order $-a$ will give a generator matrix for the code $C_L(B, aP_0)$. Thus for $a = 0$ or $a = 1$ we get just the eightfold repetition code. For $1 < a < 8$ the code has rank a as can be easily checked. It is also an easy exercise to find in each such code a code word of weight $8 - a$.

If you try to perform these calculations, you may wish to begin by following the calculation of generator and check matrices for the codes that will be presented in Example 5.7.

5.4 Parameters of function codes

PROPOSITION *Let C be an algebraic curve over the field K and let G be a function code $C_L(B, D)$ defined over C with $d(D) = a$. Then for the block length n, rank m, and minimum distance d of G,*
(a) $n = d(B)$,
(b) $m = l(D) - l(D - B)$,
(c) $d \geq n - a$.

Proof Statement (a) is obvious. Statement (b) is a direct consequence of the rank and nullity theorem. The map taking a function φ in $L(D)$ to its sequence of values on the points of B is linear. The map takes φ to the all-zero sequence

Table 5.2 Basic functions for the cubic

Function	Order	(x,y)-points 1	2	3	4	5	(u,v)-pts 6	7	8
1	0	1	1	1	1	1	1	1	1
$x/(y+1)$	-2	0	0	1	10	11	1	11	10
$y/(y+1)$	-3	11	10	0	0	0	1	1	1
$x^2/(y+1)^2$	-4	0	0	1	11	10	1	10	11
$xy/(y+1)^2$	-5	0	0	0	0	0	1	11	10
$x^3/(y+1)^3$	-6	0	0	1	1	1	1	1	1
$x^2y/(y+1)^3$	-7	0	0	0	0	0	1	10	11

if and only if $\varphi(P) = 0$ for all points of P of B, which holds if and only if $\varphi \in L(-B)$. Since B and D are disjoint $\varphi \in L(D)$ and $\varphi \in L(-B)$ is equivalent to $\varphi \in L(D - B)$.

Statement (c) follows from the degree theorem (Theorem 4.2), which implies that a non-zero function has the same number of zeros and poles when they are counted with correct multiplicities. Let φ be a non-zero function of $L(D)$ and suppose that φ has b zeros among the points of B. Then by the degree theorem,

$$0 = \sum_{P \in B} \nu_P(\varphi) d(P) + \sum_{Q \notin B} \nu_Q(\varphi) d(Q).$$

But for all points P, $\nu_P(\varphi) + D(P) \geq 0$, and for the points $P \in B$, $D(P) = 0$; hence φ has no poles at points of B, and the first sum above has value at least equal to b. Hence

$$0 \geq b + \sum_{Q \notin B} \nu_Q(\varphi) d(Q) \geq b - \sum_{Q \notin B} D(Q) d(Q).$$

As $D(P) = 0$ for all points of B,

$$\sum_{Q \notin B} D(Q) d(Q) = \sum_{P \in B} D(P) d(P) + \sum_{Q \notin B} D(Q) d(Q) = d(D) = a.$$

Hence $0 \geq b - a$, or $b \leq a$. So φ has at most a zeros in B and any non-zero code word has weight at least $n - a$. □

The estimates of the proposition are most useful if $d(D) < n$. In that case $d(D - B) < 0$, so $l(D - B) = 0$. We can use Riemann's theorem to obtain a lower bound for the rank and obtain the following corollary. These lower bounds are called the *designed* rank and minimum distance of the code.

COROLLARY *If $d(D) = a < n = d(B)$, then $C_L(B, D)$ has rank at least $a+1-\gamma$, where γ is the genus of the underlying curve, and minimum distance at least $n - a$.* □

Remark When $\gamma = 0$, the proposition states that the codes meet the Singleton bound, which is not discussed further in this book (see my book 1996). It will emerge in the exercises that classical Goppa codes are geometric Goppa codes for curves of genus 0, so the proposition establishes that classical Goppa codes (and their subclass Reed–Solomon codes) meet the Singleton bound.

EXAMPLE Table 5.3 shows the true parameters of the codes for the cubic. The parameters were found by constructing the codes explicitly using generator matrices constructed from the rows of Table 5.2. You can also use the matrices presented in Example. 5.7 which are obtained by using row operations to bring the generator matrices into standard form. The genus of the curve is 1 and the block length is $d(B) = 8$ so the conditions for the corollary are satisfied. We

Table 5.3 Parameters of function codes over the cubic

a	1	2	3	4	5	6	7
Rank	1	2	3	4	5	6	7
Min. dist.	8	6	5	4	3	2	2

can read off the dimensions $l(aP_0)$ from the functions in Table 5.2. We have $l(aP_0) = a - 1$ for $a \geq 1$ and these values correspond to the rank entries in the table. The estimate for the minimum distance given by the corollary is $8 - a$, which is sharp, except for $a = 1$ and $a = 7$.

5.5 Points of the Klein quartic

EXAMPLE In this example we shall calculate a table analogous to that of Example 5.3 for the Klein quartic over F_{16}. As shown in Example 2.11 this curve has 17 points over F_{16}, which we number as in Table 5.4

We take for D a multiple aP_0 and for B the 16 other points. In Example 4.9 we found that a basis for $L(D)$ could be found by choosing functions of the form $y^i/x^j$with $0 \leq 3i \leq 2j$. These functions have order $-(3j - i)$ at P_0.

Table 5.5 shows the values at P_i for a choice of such functions of orders down to -18. The transpose of the rows of order up to $-a$ gives a generator matrix for $C_L(B, aP_0)$. The entries in the m and d columns show the true rank and minimum distance of these codes. Notice that as a increases from 16 to 17, the rank of the code fails to increase. At that stage $l(17P_0 - B) = 1$. The rank values can be obtained by applying standard row operations to the matrix. The minimum distance values are found by searching for short code words.

5.6 Primary Goppa codes

As we have seen, the field of ordinary rational functions $K(x)$ is the algebraic function field corresponding to the irreducible polynomial $f(x, y) = y$. Geometrically, the curve in question is a straight line. However, for this case the definition of a Goppa function code does not match the definition of the classical Goppa code $\mathrm{GC}(B, g)$, which we recall here

DEFINITION Let $g(z)$ be a polynomial over K, and let $B = \{\beta_1, \ldots, \beta_n\}$ be a set of elements of K such that $g(\beta_i) \neq 0$ for $i = 1, \ldots, n$. Then the (classical)

Table 5.4 Points of the Klein quartic over F_{16}

0: $(0,0)$, 1: $(u{=}0, v{=}0)$, 2: $(w{=}0, z{=}0)$,
3: $(10,11)$, 4: $(11,10)$,
5: $(3,10)$, 6: $(5,11)$, 7: $(8,10)$, 8: $(15,11)$,
9: $(10,2)$, 10: $(11,4)$, 11: $(10,9)$, 12: $(11,14)$,
13: $(6,8)$, 14: $(13,15)$, 15: $(7,3)$, 16: $(12,5)$.
(Points are (x,y)- points except for P_1 and P_2)

Table 5.5 Basic functions for the Klein quartic

Function	Order	Points 1	2	3	4	5	6	7	8	9	10	11	12	13	14	15	16	m	d
1	0	1	1	1	1	1	1	1	1	1	1	1	1	1	1	1	1	1	16
$1/x$	−3	0	0	11	10	8	15	3	5	11	10	11	10	4	9	14	2	2	13
y/x^2	−5	0	0	1	1	2	4	9	14	13	7	12	6	7	12	6	13	3	11
$1/x^2$	−6	0	0	10	11	15	3	5	8	10	11	10	11	9	14	2	4	4	10
y^2/x^3	−7	0	1	10	11	12	6	13	7	4	9	14	2	3	5	8	15	5	9
y/x^3	−8	0	0	11	10	9	14	2	4	2	4	9	14	5	8	15	3	6	8
$1/x^3$	−9	0	0	1	1	5	8	15	3	1	1	1	1	15	3	5	8	7	7
y^2/x^4	−10	0	0	1	1	4	9	14	2	7	12	6	13	12	6	13	7	8	6
y/x^4	−11	0	0	10	11	7	12	6	13	15	3	5	8	13	7	12	6	9	5
$1/x^4$	−12	0	0	11	10	3	5	8	15	11	10	11	10	14	2	4	9	10	4
y^2/x^5	−13	0	0	11	10	11	10	11	10	3	5	8	15	2	4	9	14	11	3
y/x^5	−14	0	0	1	1	10	11	10	11	13	7	12	6	6	13	7	12	12	2
$1/x^5$	−15	0	0	10	11	1	1	1	1	10	11	10	11	10	11	10	11	13	1
y^2/x^6	−16	0	0	10	11	14	2	4	9	4	9	14	2	8	15	3	5	14	1
y/x^6	−17	0	0	11	10	6	13	7	12	2	4	9	14	1	1	1	1	14	1
$1/x^6$	−18	0	0	1	1	8	15	3	5	1	1	1	1	3	5	8	15	15	1

Goppa code $\mathrm{GC}(B,g)$ is the set of words $d \in K^n$ such that

$$s(z) = \sum_{j=1}^{n} \frac{d_j}{z - \beta_j} \equiv 0 \pmod{g(z)}.$$

The following discussion leads to a definition of primary Goppa codes that includes the classical codes as a special case. You can skip over it if you wish.

In classical terms, the values d_j are the residues of $s(z)$ at the places $z = \beta_j$. By its construction $s(z)$ has degree < 0, so $\nu_\infty(s) \geq 1$. Furthermore, using the horizon theorem, the fact that $g(z)$ divides $s(z)$ is equivalent to the condition that $\nu_Q(s) \geq \nu_Q(g)$ at every zero Q of $g(z)$. So if we let B be the divisor $\sum Q_j$ where Q_j is the place $z = \beta_j$, and we let D be the divisor with value $D(Q) = \nu_Q(g)$ for all finite places and $D(\infty) = -1$, then $s(z) \in L(B - D)$. So we could say that $\mathrm{GC}(B,g)$ consists of the sequences $(d_1, \ldots, d_n)$ where d_j is the residue of a rational function $s \in L(B - D)$. This definition can be extended to any divisor that is zero at all places $z = b_j$, which is the way Goppa himself defined his codes.

It is, moreover, easily possible to remove the use of residues. With the divisors B and D as above, let $\varphi(z) \in L(D)$. Then all the poles of $\varphi(z)$ lie among the zeros of $g(z)$, and indeed $\varphi(z)s(z)$ has no poles outside the set $\{Q_1, \ldots, Q_n\}$. By the residue theorem of classical analysis, the sum of all the residues of a rational function is 0. The reader familiar with the classical theory of residues will realize that setting $D(\infty) = -1$ ensures that the residue of $\varphi(z)s(z)dz$ at ∞ is 0 (see Exercises 6-10). Thus we have $\sum d_j\varphi(Q_j) = 0$. So the code is just the orthogonal complement of the set of words of the form $(\varphi(Q_1), \ldots, \varphi(Q_n))$ for $\varphi \in L(D)$.

DEFINITION Let K be a finite field and let C be a smooth algebraic curve defined over K. Let $\{P_1, \ldots, P_n\}$ be a set of points of C of degree 1, and let B be the divisor that is the sum of these places, $\sum P_j$. Further let D be a divisor such that $D(P_j) = 0$ for all $j = 1, \ldots, n$. The *primary Goppa code*, or Goppa *residue* code, $C_\Omega(B, D)$, is defined as the set of vectors $(d_1, \ldots, d_n)$, such that $\sum d_j \varphi(P_j) = 0$ for all $\varphi \in L(D)$.

Again B determines the set of places P_i and we shall write $P_i \in B$ to indicate that P_i is one of the selected places. If D is a non-negative multiple of a single rational point we shall call the code a *one-point code.*

Primary Goppa codes can also be represented using residues of differentials over a general curve, but we shall not need that representation. The set of differentials related to a divisor D is denoted by $\Omega(D)$ which explains the notation. Differentials and residues are treated in Chapter 12, where the residue theorem is proved and the residue representation of Goppa codes is discussed.

DEFINITION For any function φ without poles among the points P_i we shall denote the value $\sum d_j \varphi(P_j)$ by $\varphi \cdot d$, and call it the syndrome of d with respect to φ.

With this notation $C_\Omega(B, D)$ can be defined as the set of words d satisfying $\varphi \cdot d = 0$ for all $\varphi \in L(D)$. Our definition provides an easy method for producing a check matrix for a residue code.

PROPOSITION *If A is a generator matrix for the function code $C_L(B, D)$, then $A^\top$ is a check matrix for the residue code $C_\Omega(B, D)$.*

Proof By the definition of the codes, u is a code word of $C_\Omega(B, D)$ if and only if $v \cdot u = 0$ for all code words v of $C_L(B, D)$. Since the columns of A form a basis of $C_L(B, D)$, that will hold if and only if $v \cdot u = 0$ for all columns of A, in other words, if and only if $A^\top u = 0$. □

5.7 Generator and check matrices

EXAMPLE In this example I give generator and check matrices for the codes $C_L(B, aP_0)$ and $C_\Omega(B, aP_0)$ derived from the functions of Table 5.2. The curve is $x^3 + y^3 = 1$, P is the point (x=0, y=1), and B is the sum of the eight other rational points over F_4. The matrices are obtained by using row operations on appropriate rows of the following matrix extracted from Table 5.2. The matrices in the second column are found using Proposition 5.1.

$$\begin{bmatrix} 1 & 1 & 1 & 1 & 1 & 1 & 1 & 1 \\ 0 & 0 & 1 & 10 & 11 & 1 & 11 & 10 \\ 11 & 10 & 0 & 0 & 0 & 1 & 1 & 1 \\ 0 & 0 & 1 & 11 & 10 & 1 & 10 & 11 \\ 0 & 0 & 0 & 0 & 0 & 1 & 11 & 10 \\ 0 & 0 & 1 & 1 & 1 & 1 & 1 & 1 \\ 0 & 0 & 0 & 0 & 0 & 1 & 10 & 11 \end{bmatrix}$$

For the residue code $C_\Omega(B, aP_0)$, the matrix in the first column is a check matrix and the matrix in the second is a transposed generator matrix. For the function code $C_L(B, aP_0)$ it is the other way round. The matrix in the first column is a transposed generator matrix and the matrix in the second is a check matrix. I also list the block length, rank, and minimum distance of both codes (the block length is always 8).

$a = 1$, C_Ω $(n = 8, m = 7, d = 2)$, C_L $(n = 8, m = 1, d = 8)$:

$$\begin{bmatrix} 1 & 1 & 1 & 1 & 1 & 1 & 1 & 1 \end{bmatrix} \qquad \begin{bmatrix} 1 & 1 & 0 & 0 & 0 & 0 & 0 & 0 \\ 1 & 0 & 1 & 0 & 0 & 0 & 0 & 0 \\ 1 & 0 & 0 & 1 & 0 & 0 & 0 & 0 \\ 1 & 0 & 0 & 0 & 1 & 0 & 0 & 0 \\ 1 & 0 & 0 & 0 & 0 & 1 & 0 & 0 \\ 1 & 0 & 0 & 0 & 0 & 0 & 1 & 0 \\ 1 & 0 & 0 & 0 & 0 & 0 & 0 & 1 \end{bmatrix}$$

$a = 2$, C_Ω $(n = 8, m = 6, d = 2)$, C_L $(n = 8, m = 2, d = 6)$:

$$\begin{bmatrix} 1 & 1 & 0 & 11 & 10 & 0 & 10 & 11 \\ 0 & 0 & 1 & 10 & 11 & 1 & 11 & 10 \end{bmatrix} \qquad \begin{bmatrix} 1 & 1 & 0 & 0 & 0 & 0 & 0 & 0 \\ 11 & 0 & 10 & 1 & 0 & 0 & 0 & 0 \\ 10 & 0 & 11 & 0 & 1 & 0 & 0 & 0 \\ 0 & 0 & 1 & 0 & 0 & 1 & 0 & 0 \\ 10 & 0 & 11 & 0 & 0 & 0 & 1 & 0 \\ 11 & 0 & 10 & 0 & 0 & 0 & 0 & 1 \end{bmatrix}$$

$a = 3$, C_Ω $(n = 8, m = 5, d = 3)$, C_L $(n = 8, m = 3, d = 5)$:

$$\begin{bmatrix} 1 & 0 & 0 & 1 & 11 & 1 & 10 & 0 \\ 0 & 0 & 1 & 10 & 11 & 1 & 11 & 10 \\ 0 & 1 & 0 & 10 & 1 & 1 & 0 & 11 \end{bmatrix} \qquad \begin{bmatrix} 1 & 10 & 10 & 1 & 0 & 0 & 0 & 0 \\ 11 & 1 & 11 & 0 & 1 & 0 & 0 & 0 \\ 1 & 1 & 1 & 0 & 0 & 1 & 0 & 0 \\ 10 & 0 & 11 & 0 & 0 & 0 & 1 & 0 \\ 0 & 11 & 10 & 0 & 0 & 0 & 0 & 1 \end{bmatrix}$$

$a = 4$, C_Ω $(n = 8, m = 4, d = 4)$, C_L $(n = 8, m = 4, d = 4)$:

$$\begin{bmatrix} 1 & 0 & 0 & 0 & 10 & 1 & 11 & 1 \\ 0 & 0 & 1 & 0 & 1 & 1 & 1 & 0 \\ 0 & 1 & 0 & 0 & 11 & 1 & 10 & 1 \\ 0 & 0 & 0 & 1 & 1 & 0 & 1 & 1 \end{bmatrix} \qquad \begin{bmatrix} 10 & 11 & 1 & 1 & 1 & 0 & 0 & 0 \\ 1 & 1 & 1 & 0 & 0 & 1 & 0 & 0 \\ 11 & 10 & 1 & 1 & 0 & 0 & 1 & 0 \\ 1 & 1 & 0 & 1 & 0 & 0 & 0 & 1 \end{bmatrix}$$

$a = 5$, C_Ω $(n = 8, m = 3, d = 5)$ C_L $(n = 8, m = 5, d = 3)$:

$$\begin{bmatrix} 1 & 0 & 0 & 0 & 10 & 0 & 0 & 11 \\ 0 & 0 & 1 & 0 & 1 & 0 & 10 & 10 \\ 0 & 1 & 0 & 0 & 11 & 0 & 1 & 11 \\ 0 & 0 & 0 & 1 & 1 & 0 & 1 & 1 \\ 0 & 0 & 0 & 0 & 0 & 1 & 11 & 10 \end{bmatrix} \qquad \begin{bmatrix} 10 & 11 & 1 & 1 & 1 & 0 & 0 & 0 \\ 0 & 1 & 10 & 1 & 0 & 11 & 1 & 0 \\ 11 & 11 & 10 & 1 & 0 & 10 & 0 & 1 \end{bmatrix}$$

$a = 6$, C_Ω $(n = 8, m = 2, d = 6)$, C_L $(n = 8, m = 6, d = 2)$:

$$\begin{bmatrix} 1 & 0 & 0 & 0 & 0 & 0 & 10 & 11 \\ 0 & 0 & 1 & 0 & 0 & 0 & 11 & 10 \\ 0 & 1 & 0 & 0 & 0 & 0 & 10 & 11 \\ 0 & 0 & 0 & 1 & 0 & 0 & 0 & 1 \\ 0 & 0 & 0 & 0 & 0 & 1 & 11 & 10 \\ 0 & 0 & 0 & 0 & 1 & 0 & 1 & 0 \end{bmatrix} \qquad \begin{bmatrix} 10 & 10 & 11 & 0 & 1 & 11 & 1 & 0 \\ 11 & 11 & 10 & 1 & 0 & 10 & 0 & 1 \end{bmatrix}$$

$a = 7$, C_Ω $(n = 8, m = 1, d = 8)$, C_L $(n = 8, m = 7, d = 2)$:

$$\begin{bmatrix} 1 & 0 & 0 & 0 & 0 & 0 & 0 & 1 \\ 0 & 0 & 1 & 0 & 0 & 0 & 0 & 1 \\ 0 & 1 & 0 & 0 & 0 & 0 & 0 & 1 \\ 0 & 0 & 0 & 1 & 0 & 0 & 0 & 1 \\ 0 & 0 & 0 & 0 & 0 & 1 & 0 & 1 \\ 0 & 0 & 0 & 0 & 1 & 0 & 0 & 1 \\ 0 & 0 & 0 & 0 & 0 & 0 & 1 & 1 \end{bmatrix} \qquad \begin{bmatrix} 1 & 1 & 1 & 1 & 1 & 1 & 1 & 1 \end{bmatrix}$$

It is striking that the parameters of the function codes are just those of the residue codes in reverse order. Indeed (as you are asked to show in Exercise 3), the residue code for a and the function code for $8 - a$ are identical. This is not an accident and is explained in Chapter 12.

5.8 Parameters of residue codes

PROPOSITION *Let C be an algebraic curve of genus γ over the field K and let G be a residue code $C_\Omega(B, D)$ defined over C with $d(D) = a$.*

Assume that $2\gamma - 2 < a$. Then for the block length n, rank m, and minimum distance d of C,
(a) $n = d(B)$,
(b) $m = n - a + \gamma - 1 + l(D - B)$,
(c) $d \geq a - (2\gamma - 2)$.

The value for the rank and the lower bound for the minimum distance are called the *designed* rank and minimum distance

EXAMPLE Table 5.6 gives the parameters for the residue codes calculated in Example 5.7

Table 5.7 gives the parameters for the residue and function Goppa codes, $C_\Omega(B, aP_0)$ and $C_L(B, aP_0)$, based on the Klein quartic. As this curve has genus 3 the proposition states that the minimum distance of residue codes is at

Table 5.6 Parameters of residue codes over the cubic

a	1	2	3	4	5	6	7
Rank	7	6	5	4	3	2	1
Min. dist.	2	2	3	4	5	6	8

Table 5.7 Parameters of codes over the Klein quartic

a	0	3	5	6	7	8	9	10	11	12	13	14	15	16	17	18
Rank C_L	1	2	3	4	5	6	7	8	9	10	11	12	13	14	14	15
Min. dist.	16	13	11	10	9	8	7	6	5	4	3	2	1	1	1	1
Rank C_Ω	15	14	13	12	11	10	9	8	7	6	5	4	3	2	2	1
Min. dist.	2	2	2	2	3	4	5	6	7	8	9	10	11	13	13	16

least $a - 4$. You will observe that this bound is exact for most values on the table, but not for the values at either end.

Proof (a) $B = \sum P_j$ and $d(P_j) = 1$, by hypothesis, so $d(B) = n$, and by definition the block length of $C_\Omega(B, D)$ is the number of places in B.

(b) Let A be a generator matrix for the function code $C_L(B, D)$. Then by Proposition 5.4, $A^\top$ has rank $l(D) - l(D - B)$. Thus by the rank and nullity theorem, C has rank $n - l(D) - l(D - B)$. As $a > 2\gamma - 2$, Riemann's theorem tells us that $l(D) = a + 1 - \gamma$.

(c) Suppose that d is a non-zero code word of C and arrange the places P_j so that $d_j \neq 0$ for $j = 1, \ldots, k$, and $d_j = 0$ for $j > k$. Put $B_j = \sum_{i=1}^{j} P_j$. We shall show that the assumption that $1 \leq k < a - (2\gamma - 2)$ leads to a contradiction. For in that case we have $d(D - B_k) > 2\gamma - 2$, and so also $d(D - B_{k-1}) > 2\gamma - 2$. Hence by Riemann's theorem $l(D - B_k) = r - k + 1 - \gamma$, and $l(D - B_{k-1}) = r - k + 2 - \gamma$. Thus there exists $\varphi \in L(D - B_{k-1})$, $\varphi \notin L(D - B_k)$. That implies that $\varphi(P_j) = 0$ for $j = 1, \ldots, k - 1$, and $\varphi(P_k) \neq 0$. As $(D - B_{k-1}) \leq D$, $\varphi \in L(D)$ and $\varphi \cdot d = d_k \varphi(P_k) \neq 0$, contradicting the assumption that $d \in C$. □

Theorem 12.17 states that there exists a divisor W of degree $2\gamma - 2$ such that $C_\Omega(B, D) = C_L(B, B + W - D)$. With that theorem it is not difficult to see that the estimates of this proposition and Proposition 5.4 are the same (see Exercise 11).

Note that if $l(D - B) = d(D - B) + 1 - \gamma$, then the rank m of the code reduces to 0, so a must be chosen at most equal to $n + 2\gamma - 2$. In order for the estimate of (b) to provide information without the need to calculate $l(D - B)$, we must also have $a < n + \gamma - 1$.

5.9 Exercises

1. Calculate the minimum distance of the function codes $C_L(B, aP)$ based on $x^3 + y^3 = 1$ directly and check their ranks.
2. Construct function codes $C_L(B, aP)$ of block length 25 based on $x^5 + y^5 = 1$. Calculate their parameters.
3. Prove that for $x^3 + y^3 = 1$, the residue and function Goppa codes of the text are pairwise identical.
4. Let Q be the unique point at infinity of the Hermitian curve $y^5 + x^4 + x$ defined over F_{16} and let B be the sum of the 64 finite points (see Exercise 7 of

Chapter 2). Consider the code $C_\Omega(B, mQ)$ that has designed minimum distance 11. What is the value of m?
Use Exercise 15 to construct a basis of $L(mQ)$ and construct a table of their values at the points $(1,6)$, $(1,7)$, $(4,4)$, $(4,5)$, $(2,2)$ which all lie on the curve.
5. Construct generator matrices for the residue codes corresponding to the codes of Exercise 2.
6. Let C be the x-axis defined by $y = 0$ defined over F_{16}. Denote the point $(x{=}\beta, y{=}0)$ by (β) and the point $(u{=}0, v{=}0)$ by (∞). Let B be the divisor $\sum(\beta)$, where the sum is taken over all non-zero values β, and let $D = 6(0) - (\infty)$. Use functions $1/x$, $1/x^2$, ..., $1/x^6$ (which are a basis of $L(D)$ by Exercise 10 of Chapter 4) to produce a check matrix for $C_\Omega(B, D)$.

The next two exercises are for readers familiar with BCH codes.

7. Show that if the columns of the check matrix obtained in Exercise 6 are arranged correctly, then the matrix is the Vandermonde check matrix $V_{4,3}$ of the BCH code BCH(4,3).
8. Prove that by using the x-axis in the manner of Exercise 6, all BCH and Reed–Solomon codes can be represented as geometric Goppa codes.

The next two exercises are for readers familiar with classical Goppa codes.

9. With C as in Exercise 6, let B' be the divisor $\sum(\beta)$, where the sum is over all β. Furthermore, let Q be the place defined by the irreducible polynomial $g(z) = z^3 + z + 1$. Let D' be the divisor $Q - (\infty)$ and D'' be the divisor $2Q - (\infty)$. Using the functions obtained in Exercise 11 of Chapter 4, construct check matrices for $C_\Omega(B', D')$ and $C_\Omega(B', D'')$. Verify that these are check matrices of classical Goppa codes for g and g^2 (these codes are used as examples of classical Goppa codes in my book (Pretzel 1996)).
10. Prove that by using C as in Exercise 6, all classical Goppa codes can be obtained as geometric Goppa codes.
11. Let $B = P_1 + \cdots + P_n$ be a sum of rational points of the smooth curve C and let the divisor D be disjoint from B. Assume further that $2\gamma - 2 < d(D)$. Let W be a divisor such that $d(W) = 2\gamma - 1$ and $W(P_i) = -1$. Show that the estimates for the parameters of $C_\Omega(B, D)$ given by Proposition 5.4 and those for the parameters of $C_L(B, B + W - D)$ given by Proposition 5.8 are the same.

6

BASIC ERROR PROCESSING

The theoretical properties of a class of codes only bear fruit if there exists a practical error processing scheme that exploits them. So there have been continuing efforts to produce practical error processors for Goppa codes. The first processing algorithm was published by Justesen *et al.* (1989). It applied only to a somewhat restricted class of codes based on smooth plane curves and could not correct errors up to the full designed capability of the code. This algorithm was generalized by Skorobogatov and Vlăduţ (1990). Their algorithm removes the restriction to plane curves, but it still does not decode to the full capability of the code. Feng and Rao (1993) extended the ideas of the processor of Justesen *et al.* so that the processor could correct errors up to the designed capability of the code (and sometimes beyond that). Their clever idea was to use a majority voting scheme to guess additional syndromes. Inspired by their paper, Duursma (1993) produced an analogous extension for the Skorobogatov–Vlăduţ processor.

There have been alternative approaches to the correction problem (Porter *et al.* 1992) and a number of papers have been published that improve the speed of the algorithms by replacing the Gaussian elimination steps by more complicated but faster algorithms (Sakata *et al.* 1995). In this chapter I shall describe the Skorobogatov–Vlăduţ algorithm (or SV-algorithm for short). In the next chapter I shall introduce an extension that is a version of the Duursma algorithm (the DU-algorithm).

Throughout we fix a residue code $C_\Omega(B, D)$ based on a curve of genus γ, where $B = P_1 + \cdots + P_n$ is a sum of distinct rational points and $D(P_l) = 0$ for all l. We abuse notation by using B also to represent the set of points $\{P_1, \ldots, P_n\}$ and express the condition that $D(P_l) = 0$ for all $P_l \in B$ by saying that the *support* of D is disjoint from B. We shall assume the word c is transmitted and that the word f is received where $f = c + e$. The word e is called the *error word.*

6.1 Syndromes

DEFINITION For any word $(f_1, \ldots, f_n)$, and any function φ in the function field of the curve, we define

$$\varphi \cdot f = \sum_{l=1}^{n} \varphi(P_l) f_l,$$

and call it the *syndrome* of f with respect to φ. If φ has a pole at P_l for any l, we set $\varphi \cdot f = \infty$.

Using this notation we can say that our code consists of those words c for which $\varphi \cdot c = 0$ for all $\varphi \in L(D)$. It follows that if $f = c + e$, where c is a code word, then $\varphi \cdot f = \varphi \cdot e$ for all $\varphi \in L(D)$. We thus have a lot of information about e in the form of the syndromes of f. The task of an error processor is to determine e from the syndromes of f, assuming that the weight of e is at most t. The larger t is, the better the error processor, but t is limited by the condition that $2t < d$ where d is the minimum distance of the code. The SV-algorithm requires the existence of certain auxiliary divisors. They can always be found if $2t < d - \gamma$, but may not exist for larger values of t.

EXAMPLE In discussing the SV-algorithm I shall use the example $C_\Omega(B, aP_0)$ based on $x^3 + y^3 = 1$ from the previous chapter (see Example 5.7). The curve has genus 1.

We take c, e, and f as follows:

$$\begin{aligned} c &= 0 \quad 0 \; 10 \; 11 \quad 1 \; 10 \quad 1 \; 11 \\ e &= 1 \quad 0 \; 11 \quad 0 \quad 0 \quad 0 \quad 0 \quad 0 \\ f &= 1 \quad 0 \quad 1 \; 11 \quad 1 \; 10 \quad 1 \; 11 \end{aligned}$$

We shall need bases of $L(aP_0)$ for $a \leq 6$. These are naturally chosen from the sequence

$$\begin{array}{cccccc} 0 & 2 & 3 & 4 & 5 & 6 \\ 1 & x/(y+1) & y/(y+1) & x^2/(y+1)^2 & xy/(y+1)^2 & x^3/(y+1)^3, \end{array}$$

where the first row of the table denotes the pole order of the function at P_0. The syndromes of f are computed from the table of values in Example 5.3.

$$\begin{array}{rcrcl} 1 \cdot f & = & 1 + 0 + 1 + 11 + \;\; 1 + 10 + \;\; 1 + 11 & = & 10 \\ x/(y+1) \cdot f & = & 0 + 0 + 1 + \;\; 1 + 11 + 10 + \;\; 1 + 10 & = & 10 \\ y/(y+1) \cdot f & = & 11 + 0 + 0 + \;\; 0 + \;\; 0 + 10 + \;\; 1 + 11 & = & 11 \\ x^2/(y+1)^2 \cdot f & = & 0 + 0 + 1 + 10 + 10 + 10 + 10 + 10 & = & 11 \\ xy/(y+1)^2 \cdot f & = & 0 + 0 + 0 + \;\; 0 + \;\; 0 + 10 + 11 + \;\; 1 & = & 0 \\ y^2/(y+1)^2 \cdot f & = & 0 + 0 + 0 + \;\; 0 + \;\; 0 + 10 + 10 + 10 & = & 10 \end{array}$$

Similar calculations show that c is a code word of $C_\Omega(B, 6P_0)$ and that e has the same syndromes as f. The weight of e is 2, while the minimum distance of $C_\Omega(B, aP_0)$ is a. So e should be correctable for $a \geq 5$. In fact the SV-algorithm requires $a = 6$ and would not work if I had chosen a code word outside $C_\Omega(B, 6P_0)$.

6.2 Error locators

The algorithms I am going to describe are based on the idea of first determining (to some extent) which values e_l are non-zero and then using this information to calculate the precise error values. To that end we make the following definition.

DEFINITION For an error word $e = (e_1, \ldots, e_n)$ the point P_l is called an *error location* if $e_l \neq 0$. A non-zero function θ is called an *error locator* if $\theta(P_l) = 0$ for all error locations and θ has no poles among $P_1, \ldots, P_n$. Thus $\theta(P_l)e_l = 0$ for all $l = 1, \ldots, n$.

The following proposition shows that a good error locator determines not just the error locations, but also the values e_l at these locations.

PROPOSITION *Let e have weight at most t and let A and X be arbitrary divisors with support disjoint from B, such that $d(A) \leq t + r$ and $d(X) \geq t + r + 2\gamma - 1$. If $l(A) > t$, then there exists an error locator in $L(A)$. Whatever the dimension $l(A)$ is, e is uniquely determined by any error locator $\theta \in L(A)$ and the syndromes $\varphi \cdot e$ of e with respect to functions in $L(X)$.*

Remark If we can choose $X \leq D$, then the syndromes we require can be calculated from the received word.

Proof Let M be the support of e, that is the set $P_l \in B$ such that $e_l \neq 0$. Choose a basis $\psi_1, \ldots, \psi_s$ of $L(A)$, so $l(A) = s$. Then a linear combination $\theta = \alpha_1\psi_1 + \cdots + \alpha_s\psi_s$ is an error locator if and only if

$$\alpha_1\psi_1(P_l) + \cdots + \alpha_s\psi_s(P_l) = 0$$

for all $P_l \in M$. That is a system of at most t equations in s unknowns. It has a non-zero solution if $s > t$. Thus under the hypothesis, $L(A)$ contains an error locator.

Now suppose that θ is an error locator in $L(A)$ and let M' be the set of points $P_l \in B$ with $\theta(P_l) = 0$. It follows from the degree theorem that the number of points $P_l \in M'$ is at most $t + r$ and among these are the t true error locations of e. Thus for any word e with θ as an error locator and any function φ without poles in B we have

$$\varphi \cdot e = \sum_{P_l \in M'} \varphi(P_l)e_l. \qquad (*)$$

That holds in particular for all $\varphi \in L(X)$. If e and e' are two words such that $\varphi \cdot e = \varphi \cdot e'$ for all $\varphi \in L(X)$, then $\varphi \cdot (e - e') = 0$ for all $\varphi \in L(X)$ and so $e - e'$ is a code word of $C_\Omega(B, X)$. By Proposition 5.8 the minimum distance of this code is at least $t + r + 1$. If e and e' both have support in M', then $e - e'$ has at most $t + r$ non-zero entries. It follows that $e - e' = 0$. Therefore eqn $(*)$ has a unique solution e with θ as an error locator. □

COROLLARY *Let M' be a set of at most $t+r$ points including the error locations of e and let $\varphi_1, \ldots, \varphi_u$ be a basis of $L(X)$ where $d(X) \geq t + r + 2\gamma - 1$. Then e is the unique solution of the system of equations*

$$\varphi_i \cdot e = \sum_{l \in M'} \varphi_i(P_l)e_l,$$

for $i = 1, \ldots, u$. □

EXAMPLE We choose $A = 3P_0$ and $X = 4P_0$. The function

$$\theta = (11y + x + 1)/(y + 1) = 1 + x/(y + 1) + 10y/(y + 1)$$

is an error locator for the error word e we chose in Example 6.1 and it lies in $L(A)$. Indeed, its values at the points P_l are

$$0\ 10\ 0\ 11\ 10\ 10\ 0\ 1.$$

Evaluating eqn (*) for the basis $1, x/(y+1), y/(y+1), x^2/(y+1)$ of $L(X)$ we obtain

$$\begin{bmatrix} 1 & 1 & 1 \\ 0 & 1 & 11 \\ 11 & 0 & 1 \\ 0 & 1 & 10 \end{bmatrix} \begin{bmatrix} e_1 \\ e_3 \\ e_7 \end{bmatrix} = \begin{bmatrix} 10 \\ 11 \\ 11 \\ 11 \end{bmatrix}.$$

This has the unique solution $e_1 = 1$, $e_3 = 11$, $e_7 = 0$, giving the correct error word.

6.3 Finding an error locator

The result of the previous paragraph does not explain how to find an error locator that can be used to determine e. That problem is solved by the following proposition.

PROPOSITION *Let e have weight at most t and let Y be an arbitrary divisor with support disjoint from B such that $d(Y) \geq t + 2\gamma - 1$. Then a function θ without poles in B is an error locator if and only if $\theta\chi \cdot e = 0$ for all $\chi \in L(Y)$.*
Remark If $\theta \in L(A)$, then the function $\theta\chi$ lies in $L(Y + A)$. So if $Y + A \leq D$ the syndrome $\theta\chi \cdot e$ can be calculated from the received word.

Proof Consider the word $e' = (\theta(P_1)e_1, \ldots, \theta(P_n)e_n)$. This word has weight at most equal to the weight of e which is t and

$$\theta\chi \cdot e = \sum \chi(P_l)\theta(P_l)e_l = \chi \cdot e'.$$

The function θ is an error locator if and only if $e' = 0$. For any function χ without poles in B it is certainly true that $\chi \cdot 0 = 0$. So if θ is an error locator certainly $\theta\chi \cdot e = 0$.

Conversely, if $\chi \cdot e' = 0$ for all $\chi \in L(Y)$, then $e' \in C_\Omega(B, Y)$. By construction, $d(Y) \geq t + 2\gamma - 1$ and so the minimum distance of this code is at least $t + 1$. Hence $e' = 0$ and therefore θ is an error locator. □

COROLLARY *Let the elements ψ_j form a basis of $L(A)$ and the elements χ_k form a basis of $L(Y)$, and let S be the matrix of syndromes $(\psi_j\chi_k \cdot e)$. Then $\theta = \sum \alpha_j\psi_j$ is an error locator in $L(A)$ if and only if $\sum \alpha_j S^j = 0$, where S^j is the jth row of S.* □

EXAMPLE The condition of this proposition need only be checked for a basis of $L(Y)$. For $t = 2$ we take $A = 3P_0$ because Riemann's theorem gives us $l(3P_0) \geq 3 + 1 - \gamma = 3$. In that case $r = \gamma = 1$ and this is the reason that we must allow non-zero values of r in the propositions. We take $Y = 3P_0 = A$ so that $A + Y = D$. We choose as a basis of $L(A)$ (and $L(Y)$) the first three functions 1, $x/(y+1)$, and $y/(y+1)$ in the list above. They have been constructed so that their products also occur in the list. That is not always possible, but since the products are certainly in $L(Y + A)$ they can always be expressed in terms of a basis of that space.

A function θ of $L(A)$ can be expressed as a linear combination of the basis functions of $L(A)$, say $\theta = \alpha_1\psi_1 + \cdots + \alpha_l\psi_l$. Then for a basis function χ_k of $L(Y)$ the syndrome of $\theta\chi_k$ is also a linear combination of the syndromes of the functions $\psi_j\chi_k$ with the same coefficients:

$$\theta\chi_k \cdot e = \alpha_1\psi_1\chi_k \cdot e + \cdots + \alpha_l\psi_l\chi_k \cdot e.$$

In our case there are three functions in both bases and the syndromes of their products form a 3×3 matrix (with rows corresponding to A and columns to Y). This matrix is

$$\begin{bmatrix} 10 & 11 & 11 \\ 11 & 11 & 0 \\ 11 & 0 & 10 \end{bmatrix}.$$

The condition that a linear combination of the rows should be zero gives three equations in three unknowns:

$$\begin{aligned} 10\alpha_1 + 11\alpha_2 + 11\alpha_3 &= 0 \\ 11\alpha_1 + 11\alpha_2 \qquad\quad &= 0 \\ 11\alpha_1 \qquad\quad + 10\alpha_3 &= 0 \end{aligned}$$

with the solution

$$\alpha_1 = 1,\ \alpha_2 = 1,\ \alpha_3 = 10.$$

That corresponds to our error locator

$$\theta = 1 + x/(y+1) + 10y/(y+1) = (11y + x + 1)/(y+1).$$

The minimum distance of $C_\Omega(B, D)$ is 6, but the method does not work for $C_\Omega(B, 5P_0)$ which should also correct double errors, because we need $A = 3P_0$ to ensure the existence of an error locator and $Y = A$ so that the error locator can be found by solving linear equations. Then the product $\psi\chi$ of functions $\psi \in L(A)$ and $\chi \in L(Y)$ may not lie in $L(5P_0)$, and so we may be unable to calculate the syndrome from the received word. In particular, the syndrome $\psi_3\chi_3 \cdot e$ is $y^2/(y+1)^2 \cdot e$, and this function does not lie in $L(5P_0)$. Hence if we start with a code word of $C_\Omega(B, 5P_0)$ that does not lie in $C_\Omega(B, 6P_0)$, then $y^2/(y+1)^2 \cdot c \neq 0$. So $y^2/(y+1)^2 \cdot e \neq y^2/(y+1)^2 \cdot f$ for the received word f.

6.4 Conditions for the SV-algorithm

Let us summarize the conditions for the SV-algorithm derived from the propositions that have been proved above.

We assume that $2\gamma - 2 < d(D)$. That ensures that our code has the parameters given by Proposition 5.8.

The subsidiary divisor A we require for the error processor must satisfy the following inequalities:

(1) $l(A) > t$,

(2) $d(A) < d(D) - (2\gamma - 2) - t$.

The first inequality ensures that there is an error locator in $L(A)$ and implies that $d(A) = t + r$ with $r \geq 0$. The second ensures that we can find Y with $d(Y) = t + 2\gamma - 1$ such that $Y + A \leq D$, so that we can find the error locator by solving a set of equations.

The conditions imply that $d(D) - (2\gamma - 2) > 2t$ so the code has a sufficient minimum distance to correct t errors.

For one-point codes with $D = mQ$ it is natural to take $A = aQ$ and $Y = (t+2\gamma-1)Q$. The following easy proposition translates the conditions (1) and (2) into conditions on the numbers m and a.

PROPOSITION *Let $C = C_\Omega(B, D)$ be a one-point code with $D = mQ$ for a non-negative integer $m > 2\gamma - 2$.*
(a) *If a satisfies*

$$t + \gamma \leq a \leq -2\gamma - t + 1,$$

then the divisor $A = aQ$ satisfies the conditions for the algorithm to work.
(b) *If $2t \leq m - 3\gamma + 1$, then there exists a number a satisfying the inequality of part* (a).

For codes with designed minimum distance $d = m - 2\gamma + 2$, the algorithm corrects t errors for $2t < d - \gamma$.

Proof (a) The right hand inequality is just a direct translation of condition (2). For condition (1) we need only verify that $l(aQ) > t$. By Riemann's theorem

$$l(aQ) \geq a + 1 - \gamma > t.$$

(b) The inequality just states that $t + \gamma \leq m - 2\gamma - t + 1$. □

6.5 The Skorobogatov–Vlăduţ error processing algorithm

I shall now state the SV-algorithm formally. You will see that the calculations conform directly to the propositions we have proved. The next paragraph provides a formal verification of the algorithm and then we calculate another example.

ALGORITHM Let c be a code word of the code $C_\Omega(B, D)$ defined over a curve of genus γ and $d(D) > 2\gamma - 2$. Let $f = c + e$, where c is a code word.

Step 0 Preliminary calculations
This step is performed once only for any given code.
Choose a divisor A such that $l(A) > t$ and $d(A) < d(D) - (2\gamma - 2) - t$.
Choose divisors X and Y such that $X \le A+Y \le D$, $d(X) \ge d(A)+2\gamma-1$, and $d(Y) \ge t+2\gamma-1$; for instance, it is possible to choose $X = D$ and $Y = D - A$. Choose bases $\{\varphi_1, \ldots, \varphi_u\}$ of $L(X)$, $\{\psi_1, \ldots, \psi_s\}$ of $L(A)$, and $\{\chi_1, \ldots, \chi_r\}$ of $L(Y)$.

If $D = mQ$ for a single point Q, then A, X, and Y can all be chosen to have the form $m'Q$ and the bases can all be chosen as subsets of a single basis of $L(D)$.

Step 1 Syndrome calculation
Given a received word f calculate the syndromes $S_{jk}(f) = \psi_j\chi_k \cdot f$ for $1 \le j \le s$ and $1 \le k \le r$. If all syndromes are 0, then f is a code word.

Step 2 Find error locator
Find a non-zero solution of the set of s linear equations

$$\sum_{j=1}^{s} S_{jk}(f)\alpha_j = 0, \qquad k \le r.$$

For the solution α let $\theta = \sum \alpha_j\psi_j$. Then θ is an error locator for e in $L(A)$.

Step 3 Find error locations
Determine the set $M \subseteq B$ of P_l for which $\theta(P_l) = 0$. This set contains the error locations.

Step 4 Calculate error values
Solve the equations

$$\sum_{P_l \in M} \varphi_i(P_l)e_l = \varphi_i \cdot f, \qquad i = 1, \ldots, u.$$

Extend the solution e of this set of equations by putting $e_j = 0$ for $P_j \notin M$. Then $e = (e_1, \ldots, e_n)$ is the error word. □

6.6 Verification

The theorem below provides a formal verification of the SV-algorithm.

THEOREM *Let c be a code word of the code $C_\Omega(B, D)$ defined over a curve of genus γ with $d(D) > 2\gamma - 2$ and let $f = c + e$. Suppose that divisors A, X, and Y have been chosen satisfying the conditions of Step 0 of the SV-algorithm for the value t and that e has weight at most t.*
(a) *The equations of Step 2 have a non-zero solution.*
(b) *The function θ determined in Step 2 is a true error locator.*
(c) *The error word e is the unique solution of the equations in Step 4.*

Proof By Proposition 6.2 there exists a non-zero error locator in $L(A)$. This function is a linear combination $\sum \alpha_j \psi_j$ because the functions ψ_j form a basis of $L(A)$, and of course, not all the values α_j are 0. It follows by Proposition 6.3 that $(\sum \alpha_j \psi_j)\chi_k \cdot e = 0$ for all χ_k. As $\psi_j \chi_k \in L(D)$ we know that $(\sum \alpha_j \psi_j)\chi_k \cdot e = (\sum \alpha_j \psi_j)\chi_k \cdot f$. Hence there exists a non-zero solution of the equations of Step 2, which establishes (a). Conversely, Proposition 6.3 shows that any such solution corresponds to an error locator $\sum \alpha_j \psi_j$, so (b) is established.

Having chosen $X \leq D$, we know that $\varphi \cdot e = \varphi \cdot f$ for all $\varphi \in L(X)$. So by Proposition 6.2 again e is the unique vector for which θ is an error locator and $\varphi \cdot e = \varphi \cdot f$ for all $\varphi \in L(X)$. That is precisely the condition expressed by the equations of Step 4 and so they have a unique solution e as claimed.

□

6.7 Another example

Here is a second example of the algorithm. We take a code $C_\Omega(B, aP_0)$ based on the Klein quartic over the field F_{16}, using Tble 5.4. The code has block length 16 and we shall use it to correct three errors.

EXAMPLE As the curve has genus 3, the algorithm requires minimum distance 10 to correct three errors. From Table 5.6 we see that we must use $C_\Omega(B, 14P_0)$.

We select the code word c, error word e, and received word f as follows:

$$
\begin{aligned}
c &= 9\ 9\ 15\ 4\ 4\ 10\ 6\ 2\ 1\ 11\ 10\ 8\ 13\ 0\ 0\ 1 \\
e &= 0\ 0\ \ 0\ 0\ 0\ \ 0\ 0\ 0\ 0\ \ 0\ \ 0\ 0\ \ 0\ 3\ 2\ 1 \\
f &= 9\ 9\ 15\ 4\ 4\ 10\ 6\ 2\ 1\ 11\ 10\ 8\ 13\ 3\ 2\ 0
\end{aligned}
$$

We number the functions of Table 5.5 by their order. Then a basis of $L(14P_0)$ consists of $\varphi_0, \varphi_3, \varphi_5, \ldots, \varphi_{14}$.

φ_0	φ_3	φ_5	φ_6	φ_7	φ_8	φ_9	φ_{10}	φ_{11}	φ_{12}	φ_{13}	φ_{14}
1	$\frac{1}{x}$	$\frac{y}{x^2}$	$\frac{1}{x^2}$	$\frac{y^2}{x^3}$	$\frac{y}{x^3}$	$\frac{1}{x^3}$	$\frac{y^2}{x^4}$	$\frac{y}{x^4}$	$\frac{1}{x^4}$	$\frac{y^2}{x^5}$	$\frac{y}{x^5}$

The auxiliary divisor A is chosen as aP_0, where $3 + 3 \leq a \leq 14 - 3 - 6 + 1$. So $a = 6$. The basis functions ψ_i are the same as the functions φ_i for $i = 0, 3, 5, 6$. The divisor Y must satisfy $d(Y) \geq 3 + 5 = 8$. So we take $Y = 8P_0$. Then the basis functions are $\chi_i = \varphi_i$ for $i = 0, 3, 5, 6, 7, 8$. Finally, X must satisfy $d(X) \geq 6 + 5 = 11$ so we take $X = 11P_0$ and a basis of $L(X)$ is obtained by omitting the last three members of a basis of D.

The syndromes are given in the following table. All of them, with the exception of $S_{5,7}$, can be calculated as $\varphi_j \cdot d$. The exceptional function is y^3/x^5 which can easily be evaluated at all the points P_l. Alternatively we can substitute $y^3 = x + x^3 y$ to obtain $y^3/x^5 = 1/x^4 + y/x^2 = \varphi_{12} + \varphi_5$ and evaluate the syndrome using these functions.

$i\backslash j$	0	3	5	6	7	8
0	0	5	12	11	9	5
3	5	11	5	7	14	14
5	12	5	14	14	11	9
6	11	7	14	7	9	12

The table is symmetric because we are dealing with a one-point code. In general that would not be the case, but it simplifies the calculations. Our first set of equations is

$$\begin{bmatrix} 0 & 5 & 12 & 11 \\ 5 & 11 & 5 & 7 \\ 12 & 5 & 14 & 14 \\ 11 & 7 & 14 & 7 \\ 9 & 14 & 11 & 9 \\ 5 & 14 & 9 & 12 \end{bmatrix} \begin{bmatrix} \alpha_1 \\ \alpha_2 \\ \alpha_3 \\ \alpha_4 \end{bmatrix} = \begin{bmatrix} 0 \\ 0 \\ 0 \\ 0 \\ 0 \\ 0 \end{bmatrix}.$$

A non-zero solution is $\alpha_1 = 11$, $\alpha_2 = 5$, $\alpha_3 = 6$, $\alpha_4 = 1$. The corresponding error locator function $11\varphi_0 + 5\varphi_3 + 6\varphi_5 + \varphi_6$ has values

$$11\ 11\ 11\ 15\ 12\ 10\ 12\ 14\ 2\ 5\ 12\ 13\ 4\ 0\ 0\ 0.$$

It follows that the set of error locations M consists of the last three points of B:

$$M = \{(13, 15), (7, 3), (12, 5)\}.$$

We use Table 5.5 to determine the matrix of the second set of equations. It consists of the last three columns of that table down to order -11. The constants on the right hand side are the syndromes of the functions of order down to -11 which can be read from the syndrome table. Therefore the second set of equations is

$$\begin{bmatrix} 1 & 1 & 1 \\ 9 & 14 & 2 \\ 12 & 6 & 13 \\ 14 & 2 & 4 \\ 5 & 8 & 15 \\ 8 & 15 & 3 \\ 3 & 5 & 8 \\ 6 & 13 & 7 \\ 7 & 12 & 6 \end{bmatrix} \begin{bmatrix} \alpha_1 \\ \alpha_2 \\ \alpha_3 \end{bmatrix} = \begin{bmatrix} 0 \\ 5 \\ 12 \\ 11 \\ 9 \\ 5 \\ 7 \\ 14 \\ 14 \\ 7 \end{bmatrix}$$

which are of course dependent. The first three equations are already sufficient to determine $\alpha_1 = 3$, $\alpha_2 = 2$, $\alpha_3 = 1$, but the rest confirm that this solution is valid. Thus we obtain

$$\begin{aligned} e &= 0\,0\ 0\,0\,0\ 0\,0\,0\,0\ 0\ 0\,0\ 0\,3\,2\,1; \\ c = f - e &= 9\,9\,15\,4\,4\,10\,6\,2\,1\,11\,10\,8\,13\,0\,0\,1. \end{aligned}$$

6.8 A comparison with Reed–Solomon codes

For readers familiar with Reed–Solomon codes I include a comparison of our examples with these standard codes.

EXAMPLE The example code based on the Klein quartic is clearly inferior to the triple error-correcting Reed–Solomon code RS(4,3) defined over F_{16}, which has block length 15, rank 9, and minimum distance 7. But the performance of geometric codes improves dramatically when the code is based on a curve with many rational points. To give an indication of this, consider the curve $x^5 + y^5 = 1$ over F_{16} which has 65 rational points. Choose one of them, P, and consider the one-point code with B as the sum of the other 64 points and $D = 37P$. By the Plücker formula, the curve has genus 6, and so this code has rank at least 32 and minimum distance $d \geq 27$. The SV-algorithm can correct 10 errors.

For comparison, consider the quadruple error-correcting code RS(4,4) defined over F_{16} which has block length 15, rank 7, and minimum distance 9. We compare the error probabilities for transmission of a code word of the Goppa code and four code words of the Reed–Solomon code. This is biased in favour of the Reed–Solomon code, as it has a poorer rate because four of its code words only transmit 28 message symbols, whereas a code word of the Goppa code transmits 32.

The probability that a Reed–Solomon code word has at most four errors is

$$q = \sum_{i=0}^{4} \binom{15}{i} p^i (1-p)^{15-i}$$

and the probability that an amalgam of four words is correctable is q^4. On the other hand the probability that a code word of the Goppa code has at most 10 errors is

$$r = \sum_{i=0}^{10} \binom{64}{i} p^i (1-p)^{64-i}.$$

The value of r is greater than q^4 for $p < 9.8\%$. Taking $p = 1\%$, we find that the probability of an uncorrectable pattern in four Reed–Solomon code words is approximately 10^{-5} while the probability of an SV-uncorrectable error pattern in a code word of the Goppa code is approximately 10^{-10}. For a message of 10000 words the chance of an uncorrectable error with the Reed–Solomon system is almost 10%, while with the Goppa code it is less than 0.00001%. If we exploit the full error-correcting capability of the Goppa code the difference will be even more marked.

6.9 Exercises

1. Use the Skorobogatov–Vlăduţ error processor for the code $C_\Omega(B, 14P_0)$ based on the Klein quartic to correct

9 2 15 4 6 10 6 2 1 12 10 8 13 0 0 1.

2. Let C be the x-axis defined by $y = 0$ over F_{16}. For any $\beta \in \mathsf{F}_{16}$ let (β) denote the point $(x{=}\beta, y{=}0)$ by and (∞) denote the point $(u{=}0, v{=}0)$. Let

B be the divisor $\sum(\beta)$, where the sum is taken over all non-zero values β, let $D = 6(0) - (\infty)$, and let A be the divisor $3(0)$. Show that A satisfies the conditions for the SV error processor for $C_\Omega(B, D) = \text{RS}(4,3)$. Use the functions $1, 1/x, 1/x^2, \ldots, 1/x^3$ as a basis of $L(A)$, the functions $1/x, \ldots, 1/x^3$ as a basis of $L(D - A)$, and the functions $1/x, \ldots, 1/x^6$ as a basis of $L(D)$ and the SV error processor to correct the word

$$d = 14\ 3\ 8\ 14\ 3\ 8\ 5\ 11\ 6\ 9\ 9\ 14\ 3\ 13\ 6.$$

3. Let C be the geometric Goppa code $C_\Omega(B, D)$ based on the curve $x^5 + y^4 + y = 0$ where B is the sum of all finite points and $D = 29Q$, where Q is the unique point at infinity. What are the parameters of this code?

Suppose the error vector of a transmitted word has the following non-zero components (and no others):

Point	$(4,4)$	$(9,9)$	$(14,14)$	$(2,2)$	$(0,0)$	$(1,6)$
Error	15	7	2	12	3	3

Calculate the syndrome table using the auxiliary divisor $A = 12Q$ and the bases of the $x^i y^j$ with $0 \le i \le 4$. Using this syndrome table calculate the error vector using the SV-algorithm.

You should, of course, get the original vector as your answer.

4. If you have read Paragraph 6.8, calculate the error probabilities of the example codes given there.

The numbers involved are very close to 1, so you will need a high precision calculator (at least 16 decimal places). If you do not have such a calculator replace the error probability by 0.005. That produces values that can be calculated with an eight-digit scientific calculator.

The next two exercises are for readers familiar with the Peterson–Gorenstein–Zierler (PGZ) error processor.

5. Compare the calculations of Exercise 2 with those of the PGZ-processor for RS(4,3) applied to the same code.

6. Prove that when applied to Reed–Solomon or BCH codes the SV error processor is the same as the PGZ error processor.

7

FULL ERROR PROCESSING

The SV error processor is quite simple, but it does not exploit the full correction capacity t of the code it is designed for. It only corrects up to $t - \gamma/2$ errors for a code based on a curve of genus γ. This weakness is quite serious because the best geometric Goppa codes necessarily come from curves with a large genus.

Duursma's algorithm (Duursma 1993), which I shall refer to as the DU-algorithm, uses the ideas of Feng and Rao (1993) to extend the SV-algorithm in such a way that it decodes to the full designed capability of the code.

7.1 Auxiliary divisors

From now on we assume that a code $C_\Omega(B, D)$ has been used, where $B = P_1 + \cdots + P_n$ for certain rational points and that a further rational point Q exists that is distinct from all the points P_l. Such a point can always be found by extending the base field if necessary. We assume that $d(D) \geq 2\gamma + 2t - 1$ and wish to correct errors of weight up to t. For that purpose we replace D by $D' = D - (d(D) - 2\gamma - 2t + 1)Q$ so that $D' \leq D$ and $d(D') = d = 2g + 2t - 1$. The code $C_\Omega(B, D')$ has designed minimum distance $2t + 1$ and contains the original code. Using D' allows us to choose how much additional error correction over the basic SV-algorithm we wish to implement.

Choose an auxiliary divisor A with $d(A) = t$ and support disjoint from B, for example $A = tQ$. We consider the divisors $A + rQ$ with $r \leq 2\gamma - 1$. At the same time we consider the divisor $A' = D' - A - (2\gamma - 1)Q$, which also has degree $d(A') = t$. Again we consider the divisors $A' + rQ$ with $r \leq 2\gamma - 1$. Our aim is to find an error locator in $L(A + mQ)$ or $L(A' + mQ)$ with m as small as possible. Proposition 6.2 tells us that as $l(A + \gamma Q) \geq t + 1$ there will be an error locator in $L(A + \gamma Q)$. Proposition 6.3 tells us that we can find the error locator using the syndromes for products of functions in $L(A + \gamma Q)$ and $L(A' + (2\gamma - 1)Q)$.

For convenience we introduce the following notation.

DEFINITION Fix a rational point Q. For any divisor H and any non-zero function θ we denote by $\mu_{H,Q}(\theta)$ the smallest integer m such that θ lies in $L(H + (m - d(H))Q)$ assuming that such an integer exists. Otherwise we put $\mu_{H,Q}(\theta) = \infty$. We call this value the *H-order* of θ with respect to Q.

In the whole of this chapter the point Q will remain unchanged. So we omit Q and write $\mu_H(\theta)$.

PROPOSITION (a) *Let* $\mu_H(\theta) = r < \infty$. *Then* $r \geq 0$ *and* $\nu_Q(\theta) = d(H) - H(Q) - r$.

(b) *Let* $H < K$ *and* $\mu_H(\theta) = r < \infty$. *Then* $\mu_K(\theta) = r + d(K - H) - (K - H)Q$.
(c) *Let* $\mu_H(\theta) = r < \infty$ *and* $\mu_{H'}\theta' = r' < \infty$. *Then* $\mu_{H+H'}(\theta\theta') = r + r'$.
Proof (a) Let $d(H) = d$. Then the degree of $H - (d+1)Q$ is negative and so $L(H - (d+1)Q) = \{0\}$. Therefore $\mu_H(\theta) > -d - 1 + d = -1$. The hypothesis $\mu_H(\theta) = r$ implies that $\theta \in L(H + (r-d)Q) \setminus L(H + (r - d - 1)Q)$. So $\nu_Q(\theta) = -(H(Q) + r - d)$.

(b) If $\theta \in L(H + kQ)$ then certainly $\theta \in L(K + kQ)$. So $\mu_K(\theta) < \infty$. Therefore we can apply part (a) and obtain

$$\nu_Q(\theta) = d(H) - H(Q) - r = d(K) - K(Q) - \mu_K(\theta),$$

from which the result follows.

(c) Certainly $\theta\theta' \in L(H + H' + (r + r' - d(H + H'))Q)$ and $\nu_Q(\theta\theta') = \nu_Q(\theta) + \nu_Q(\theta')$, from which $\mu_{H+H'}(\theta\theta') = r + r'$ follows. □

COROLLARY *Let* $\psi_1, \ldots, \psi_m$ *be functions such that*

$$\mu_H(\psi_1) < \cdots < \mu_H(\psi_m)$$

is a complete list of the possible values $\mu_H(\psi) \leq d(H) + r$. *Then* $\psi_1, \ldots, \psi_m$ *form a basis of* $H + rQ$. □

EXAMPLE For the DU-algorithm we shall use this proposition to construct special bases φ_i of $L(A + A' + (3\gamma - 1)Q)$, ψ_j of $L(A + (2\gamma - 1)Q)$, and χ_k of $L(A' + (2\gamma - 1)Q)$, indexing the functions by their A-, A'-, or $A + A'$-orders.

As our example we take the Klein quartic and use Table 5.4. Let $Q = (0, 0)$, let B be the sum of the remaining points over F_{16} in the order of the table, and consider the one-point code $C_\Omega(B, 11Q)$.

The genus of the Klein quartic is 3, so the code has block length 16, rank 7, and minimum distance 7. We have $t = \gamma = 3$, $d(D) = 11$, and so $D' = D$.

We choose $A = 3Q$. Then $A' = A$. We shall need a basis of $L(A + A' + (3\gamma - 1)Q) = L(14Q)$, which we can read off from Table 5.5. Because the divisors are all multiples of Q, the bases can all be chosen as subsets of this common basis, and we have $\mu_A(\varphi) = \mu_{A+A'}(\varphi) = -\nu_Q(\varphi)$.

i	0	3	5	6	7	8	9	10	11	12	13	14
φ_i	1	$1/x$	y/x^2	$1/x^2$	y^2/x^3	y/x^3	$1/x^3$	y^2/x^4	y/x^4	$1/x^4$	y^2/x^5	y/x^5

The basis for $L(A+kQ)$ consists of the functions φ_i with $i \leq k+3$. In particular, the basis of $L(A) = L(A')$ consists of the first two functions.

7.2 The syndrome table

Suppose now that a word c of the code $GG(B, D)$ is transmitted and undergoes an error e of weight $\leq t$ so that the word $f = c + e$ is received. As usual, given a function φ without poles among the points P_l and a word $a = (a_1, \ldots, a_n)$ we denote the value $\sum_{l=1}^{n} \varphi(P_l)a_l$ by $\varphi{\cdot}a$. Using the special bases of $L(A+(2\gamma+1)Q)$ and $L(A' + (2\gamma + 1)Q)$ we defined above, we construct a $(t + \gamma) \times (t + \gamma)$ matrix S by defining

$$S_{jk} = \psi_j\chi_k \cdot e.$$

We index the rows and columns by the orders of the basis functions ψ_j and χ_k respectively. We call these indices the *orders* of the rows and columns. The order of the (j, k)-entry is $j + k$. Some of the entries of S are *unknown* since we do not know e, but all entries of order at most $d(D)$ are known, because the corresponding functions $\psi_j\chi_k$ lie in $L(D)$ and so

$$S_{jk} = \psi_j\chi_k \cdot f.$$

In particular, all the entries in rows and columns of order at most t are known (because the corresponding products lie in $L(D') \subseteq L(D)$). On the other hand any row of order $i > t + d(D) - d(D')$ contains an entry of order $> d(D)$ which is initially unknown. We shall call a row or column *known* if all its entries are known.

If $d(D) \geq 3\gamma + 2t - 1$, then the first t rows of our syndrome table are known and that may sometimes also happen when $d(D) < 3\gamma+2t-1$. In that case these rows form the syndrome table used by the SV-processor with auxiliary divisors $A + rQ$ and $A' + (2\gamma - 1)Q$ where $r + d$ is the order of the tth row. Then there is no need for any further complication. We can use the simple SV-algorithm of the last chapter.

Normally, however, not all these rows are known. As a first preparation for the algorithm we shall employ in that case, we express the products $\psi_j\chi_k$ corresponding to unknown entries of order up to $3\gamma + 2t - 1$ in terms of the basis φ_i of $L(A + A' + (3\gamma - 1)Q)$.

We shall denote the submatrix of rows of order up to r and columns of order up to r' by $S|_{r,r'}$.

These calculations will form Steps 0 and 1 of the algorithm.

EXAMPLE We shall use the same error word as for the SV-error processor, but change the code word so that it does not lie in $GG(B, 14Q)$.

$$\begin{array}{rcl} c &=& 8\ 0\ 3\ 11\ 15\ 3\ 1\ 9\ 15\ 5\ 0\ 14\ 1\ 1\ 1\ 1 \\ e &=& 0\ 0\ 0\ \ 0\ \ 0\ 0\ 0\ 0\ \ 0\ 0\ 0\ \ 0\ 0\ 3\ 2\ 1 \\ d &=& 8\ 0\ 3\ 11\ 15\ 3\ 1\ 9\ 15\ 5\ 0\ 14\ 1\ 2\ 3\ 0 \end{array}$$

In the syndrome table below the initial entries of the rows and columns are the basis functions and the second entries are their orders. The known syndromes are the same as the ones used in Example 6.7, so they do not need to be recalculated. The unknown products $\psi_j\chi_k$ of order up to 14 are expressed in terms of the basis functions $\varphi_0, \ldots, \varphi_{14}$.

		1	$1/x$	y/x^2	$1/x^2$	y^2/x^3	y/x^3
		0	3	5	6	7	8
1	0	0	5	12	11	9	5
$1/x$	3	5	11	5	7	14	14
y/x^2	5	12	5	14	14	$\varphi_{12}+\varphi_5$	φ_{13}
$1/x^2$	6	11	7	14	φ_{12}	φ_{13}	φ_{14}
y^2/x^3	7	9	14	$\varphi_{12}+\varphi_5$	φ_{13}	$\varphi_{14}+\varphi_7$	$*$
y/x^3	8	5	14	φ_{13}	φ_{14}	$*$	$*$

With two exceptions the unknown products yield basis functions directly. For the exceptions that do not yield a basis function the substitution $y^3 = x + x^3y$ produces the required formula:

$$\frac{y}{x^2} \cdot \frac{y^2}{x^3} = \frac{y^3}{x^5} = \frac{1}{x^4} + \frac{y}{x^2},$$

$$\frac{y^2}{x^3} \cdot \frac{y^2}{x^3} = \frac{y^4}{x^6} = \frac{y}{x^5} + \frac{y^2}{x^3}.$$

Higher order entries, which are not needed by the algorithm are represented by the symbol $*$.

The algorithm proceeds by finding an error locator from the known rows or columns if there is one, or using the fact that no error locator can be calculated from the known rows or columns to determine some of the unknown entries. This is done by a democratic voting procedure, which works because of some subtle properties of the true syndrome table.

We start by proving these theoretical results. So for the next two paragraphs we shall be dealing with the syndromes of the error word e and disregard the fact that not all syndromes are known.

7.3 Existence of error locators

Recall that our aim is to find an error locator θ with $\mu_A(\theta) = u$ or $\mu_{A'}(\theta) = u$ and u as small as possible. The following proposition, a simple application of the principle of inclusion and exclusion, assures us that as u increases the chance of finding an error locator increases very strongly.

PROPOSITION *Let e have weight at most $t = d(A) = d(A')$ and let $t \le u \le \gamma + t$. Suppose that there are no error locators θ, η for e with $\mu_A(\theta) < u$ or $\mu_{A'}(\eta) < u$. For r between u and $2\gamma - 1 + t$ denote by r' the value $2\gamma - 1 + t + u - r$. Let p be the number of values r between u and $2\gamma - 1 + t$ such that there exist error locators θ and η with $\mu_A(\theta) = r$ and $\mu_{A'}(\eta) = r'$, and let q be the number of values r for which there is neither an error locator θ with $\mu_A(\theta) = r$ nor an error locator η with $\mu_{A'}(\eta) = r'$. Then $p - q \ge u - t$.*

Proof Notice that as r ranges from u to $2\gamma - 1 + t$, r' traverses the same range backwards and the sum $r + r' = 2\gamma - 1 + t + u$ is constant. Let J be the set of indices $\{r : u \le r \le 2\gamma - 1 + t\}$, let I be the subset of those r for which there exists an error locator θ with $\mu_A(\theta) = r$, and let I' be the subset of those r for which there exists an error locator η with $\mu_{A'}(\eta) = r'$. Then $p = |I \cap I'|$ and

$$q = |J \setminus (I \cup I')| = 2\gamma + t - u - |I \cup I'|.$$

Hence $p - q = |I \cap I'| + |I \cup I'| + u - t - 2\gamma$.

By inclusion and exclusion $|I \cap I'| + |I \cup I'| = |I| + |I'|$ so we estimate $|I|$ and $|I'|$. Let $B_e = \sum_{e_l \neq 0} P_l$. Then an error locator in $L(A + (r-t)Q)$ is a non-zero element of $L(A + (r-t)Q - B_e)$ and $d(A + (2\gamma - 1)Q - B_e) \geq 2\gamma - 1$. So $l(A + (2\gamma - 1)Q - B_e) \geq \gamma$ and by Corollary 7.1 we can choose a basis of $L(A+(2\gamma-1)Q-B_e)$ by choosing elements θ_i with distinct values $\mu_{A-B_e}(\theta_i) = i$ for $i \leq 2\gamma - 1 + t - d(B_e)$. The basis contains at least γ elements θ_i and by hypothesis $\mu_A(\theta_i) \geq u$. Since $B_e(Q) = 0$, Proposition 7.1b tells us that $\mu_A(\theta_i) = i + d(B_e) \leq 2g - 1 + t$. Thus $|I| \geq \gamma$. Similarly $|I'| \geq \gamma$ and so it follows that

$$p - q = |I| + |I'| + u - t - 2\gamma \geq u - t. \qquad \square$$

EXAMPLE Continuing the example of Paragraph 7.2, there are no error locators θ for e with $\mu_A(\theta) \leq 5$, but there are error locators with larger values of μ, for instance the following normalized error locators:

$$\begin{array}{ll} \varphi_6 + 11\varphi_0 + 5\varphi_3 + 6\varphi_5 & (\mu = 6); \\ \varphi_7 + \varphi_3 + \varphi_5 & (\mu = 7); \\ \varphi_8 + \varphi_0 + \varphi_3 & (\mu = 8). \end{array}$$

As we are dealing with a one-point code and using the standard divisors, we have $A = A'$. In the table below I list the pairs r, r' for $u = 3, 4, 5, 6$. Those pairs for which an error locator with $\mu_A = r$ and also an error locator with $\mu_A = r'$ exist are marked with a $*$. There are no pairs for which there is neither an error locator with $\mu_A = r$, nor one with $\mu_A = r'$. That is because the values for which there are error locators form a sequence from $t + \gamma$ to $t + 2\gamma - 1$. In general, there can be gaps in the sequence and for sequences with gaps the value q may become non-zero (see Exercise 6).

u	p	q	(r, r')					
3	0	0	(3,8)	(4,7)	(5,6)	(6,5)	(7,4)	(8,3)
4	1	0	(4,8)	(5,7)	(6,6)*	(7,5)	(8,4)	
5	2	0	(5,8)	(6,7)*	(7,6)*	(8,5)		
6	3	0	(6,8)*	(7,7)*	(8,6)*			

The example shows that the result of the theorem is really just about the incidences between two sequences of γ of more numbers in an interval of length at most 2γ.

7.4 Testing for error locators

We also need some tests to determine possible error locators. We still assume that all entries of the syndrome table are known and try to find error locators using only some of the information that is available. We shall call our potential error locators 'candidates'. There are two types of candidates: those that correspond to linear relations between the rows of S, which we shall call row candidates, and column candidates, which correspond to relations between the columns of S.

DEFINITION A function θ is called a *row candidate* of order (r, r') if $\mu_A(\theta) = r$ and $\theta\eta \cdot e = 0$ for all functions η with $\mu_{A'}(\eta) \leq r'$.

Dually a function η is called a *column candidate* of order (r, r') if $\mu_{A'}(\theta) = r'$ and $\theta\eta \cdot e = 0$ for all functions θ such that $\mu_A(\theta) \leq r$.

We shall call θ *normalized* if it has the form $\psi_r + \theta'$ with $\mu_A(\theta') < r$. Similarly η is *normalized* if it has the form $\chi_{r'} + \eta'$ with $\mu_{A'}(\eta') < r'$.

Every candidate can be normalized by multiplying it by a suitable non-zero scalar and obviously a row candidate of order (r, r') is also a row candidate of order $(r, r' - 1)$. However, the first index r cannot be changed. The following proposition lays out the relations between candidates and the syndrome matrix.

PROPOSITION (a) *A function $\theta = \sum_{j \leq r} \alpha_j \psi_j$ is a row candidate of order r, r' if and only if $a_r \neq 0$ and the rows $S^1, \ldots, S^r$ of the partial syndrome matrix $S|_{r,r'}$ satisfy the linear equation $\sum \alpha_j S^j = 0$. The corresponding statement holds for column candidates.*

(b) *A function θ with $\mu_A(\theta) = r$ is an error locator for e if and only if it is a row candidate of order $(r, t + 2\gamma - 1)$.*

(c) *If θ is a normalized row candidate of order (r, r') and η is a normalized column candidate of order (r, r'), then*

$$\theta\chi_{r'} \cdot e = \psi_r \eta \cdot e.$$

(d) *If θ_1 and θ_2 are distinct normalized row candidates of order $(r, r' - 1)$ and*

$$\theta_1\chi_{r'} \cdot e \neq \theta_2\chi_{r'} \cdot e,$$

then there is no column candidate of order $(r - 1, r')$.

Proof (a) The condition on the rows of $S|_{r,r'}$ is equivalent to the statement that

$$\sum_{j \leq r} \alpha_j \psi_j \chi_k \cdot e = 0$$

for all $k \leq r'$. That is the same as stating that $\theta\chi_k \cdot e = 0$ for all $k \leq r'$. As the functions χ_k for $k \leq r'$ form a basis of $L(A' + (r' - t)Q)$, the condition holds if and only if $\theta\eta \cdot e = 0$ for all $\eta \in L(A' + (r' - t)Q)$, which is by definition the set of η with $\mu_{A'}(\eta) \leq r'$. Thus θ is a row candidate of order r, r' if and only if the condition holds together with the requirement that $\mu_A(\theta) = r$.

(b) This is a direct consequence of Proposition 6.3 with $Y = A' + (2\gamma - 1)Q$ which has the required degree $t + 2\gamma - 1$.

(c) Set

$$\theta = \psi_r + \theta',$$
$$\eta = \chi_{r'} + \eta',$$

with $\mu_A(\theta') < r$ and $\mu_{A'}(\eta') < r'$. Then using these formulae and the candidate properties of θ and η, we obtain the following sequence of equations:

$$\theta\chi_{r'} \cdot e = \theta\eta \cdot e - \theta\eta' \cdot e \quad = \theta\eta \cdot e$$
$$= \psi_r\eta \cdot e + \theta'\eta \cdot e = \psi_r\eta \cdot e.$$

(d) If there were such a column candidate η of order (r, r') we can assume that η is normalized. The part (c) tells us that

$$\theta_1\chi_{r'} \cdot e = \psi_r\eta \cdot e = \theta_2\chi_{r'} \cdot e,$$

contradicting the hypothesis. □

EXAMPLE As an example of this proposition, consider the partial matrix of known values of Paragraph 7.2

	0	3	5	6
0	0	5	12	11
3	5	11	5	7
5	12	5	14	14
6	11	7	14	
7	9	14		

where the first row and column give the order and the unknown entries have been left blank.

Take first the S_{66} entry. There is a unique row candidate, $\psi_6 + 6\psi_5 + 5\psi_3 + 11\psi_0$, which gives the estimate $11 \cdot 11 + 5 \cdot 7 + 6 \cdot 14 = 7$ for $S_{66} = \psi_6\chi_6 \cdot e$. By the symmetry of the matrix and the fact that $\psi_i = \chi_i$, the same statement holds for the column candidate for S_{66}.

Now take the S_{75} entry. There are several row candidates, for instance $\psi_7 + 10\psi_3 + 3\psi_0$ which gives the estimate $S_{75} = 14$, $\psi_7 + \psi_3 + \psi_2$, which gives $S_{75} = 11$, and $\psi_5 + 8\psi_4 + 14\psi_3$, which gives $S_{75} = 9$. As claimed by part (d) of the proposition, there are no column candidates.

7.5 Consistency

The first point to note about unknown entries is that if all entries of order up to $s - 1$ are known then any entry of order s determines all the others.

PROPOSITION *Suppose that there is a table entry of order s, and that the syndromes $\varphi' \cdot e$ for functions φ' with $\mu_{A+A'}(\varphi') < s$ are known. Then the value of any entry of order s determines the syndrome $\varphi_s \cdot e$ from the chosen basis of $L(A + A' + 3\gamma - 1)$, and that value determines all the entries of order s.*

Proof Let the entry be $S_{rr'}$, where $r + r' = s$. Then $S_{rr'} = \psi_r \chi_{r'} \cdot e$, and by Proposition 7.1, $\mu_{A+A'}(\psi_r \chi_{r'}) = s$. Therefore $\psi_r \chi_{r'} = \alpha \varphi_s + \varphi'$, where $\mu(\varphi') < s$, and $\alpha \neq 0$. By hypothesis $\varphi' \cdot e$ is known and thus each of the values $\psi_r \chi_{r'} \cdot e$ and $\varphi_s \cdot e$ determines the other. □

EXAMPLE Continuing the example of the previous paragraph, notice that in the table in Paragraph 7.2 the S_{66} entry is $\varphi_{12} \cdot e$. So the estimate 7 we obtained is a direct guess for the value of the syndrome of φ_{12}. On the other hand the S_{75} entry of the table in Paragraph 7.2 is $\varphi_{12} + \varphi_5$. The syndrome $\varphi_5 \cdot e$ is known: it is 12, so our three estimates for $\varphi_{12} \cdot e$ from the row candidates would be $14 + 12 = 2$, $11 + 12 = 7$, $9 + 12 = 5$.

From our knowledge of e we can check that the estimate given by S_{66} is correct, but this is not guaranteed merely by the fact that we have both a column candidate and a row candidate. As we shall see in the next paragraph, however, it follows from Proposition 7.3 that the majority of such guesses will be correct. The guesses from the row candidates for S_{75} are inconsistent and will be discarded. The fact that one of them is correct appears to be an accident. We do not have a means to distinguish it from the incorrect guesses.

7.6 Majority voting

It is time to explain how to determine additional entries of the syndrome table if no error locator can be found from the known entries.

Suppose that the known syndromes $\varphi_r \cdot e$ are those of order up to $s - 1$, so that $s \geq 2t + 2\gamma$. These determine all the table entries of order up to $s - 1$. By Proposition 7.4b, any non-trivial linear relation between the known rows (or columns) of the matrix determines an error locator. We assume that there are no such relations and consider all the entries $S_{rr'}$ with $r + r' = s$. From the known entries in the syndrome table we can determine whether there are any row candidates θ of order $(r, r' - 1)$ and any column candidates η of order $(r - 1, r')$. We shall call the entry $S_{rr'}$ a *test* entry if candidates of both types exist. For each test entry choose normalized candidates θ and η. Then by Proposition 7.4c the true value $\theta \chi_{r'} \cdot e = \psi_r \eta \cdot e$ is independent of the candidates chosen. Hence, if there exists a row candidate of order (r, r') we can assume that we have chosen θ to be that candidate and to be normalized. For such a θ we have $\theta \chi_{r'} \cdot e = 0$.

Setting $\theta \chi_{r'} \cdot e = 0$ determines the value $\psi_r \chi_{r'} \cdot e$. We write $\theta = \psi_r + \theta'$ with $\mu_A(\theta') < r$ and put $\psi_r \chi_{r'} \cdot e = -\theta' \chi_{r'} \cdot e$, where the right hand side is known, because $\mu_{A+A'} \theta' \chi_{r'} < r + r' = s$. We can also determine $\psi_r \chi_{r'} \cdot e$ using η. That always gives the same value (see Exercise 8). From the syndrome $\psi_r \chi_{r'} \cdot e$ that follows from the guess $\theta \chi_{r'} \cdot e = 0$, we determine a value for the syndrome $\varphi_s \cdot e$, which we call the *vote* of the entry $S_{rr'}$. It is a remarkable fact that the majority of votes are correct.

PROPOSITION *The number of test entries $S_{rr'}$ with $r+r'=s$ producing correct votes for the value $\varphi_s \cdot e$ exceeds those producing incorrect votes by at least $s-2\gamma-2t+1>0$. In particular there is at least one such test entry.*

Proof Let $u=s-2\gamma-t+1$. It has already been observed that the existence of an unknown entry of order s implies that $s \geq 2\gamma+2t$ and so $u>t$. All the entries in rows or columns of order up to $u-1$ are known and there are no error locators θ, η with $\mu_A(\theta)<u$ or $\mu_{A'}(\eta)<u$. The numbers $r \geq u$, $r' \geq u$ with $r+r'=s$ span the interval $u \leq r, r' \leq u+2\gamma+t-1$. Thus u, r, and r' satisfy the conditions of Proposition 7.3.

Suppose the vote from $S_{rr'}$ is incorrect; then the assumption that $\theta\chi_{r'} \cdot e = 0$ is wrong. If there were an error locator θ' with $\mu_A(\theta')=r$ then Proposition 7.4c shows that $\theta\chi_{r'} \cdot e = \theta'\chi_{r'} \cdot e$, but for an error locator θ' the syndrome $\theta'\chi_{r'} \cdot e = 0$. Thus there can be no error locator θ' with $\mu_A(\theta')=r$. Similarly there can be no error locator η' with $\mu_{A'}(\eta')=r'$. It follows that the number of incorrect votes is at most equal to the number q of pairs (r,r') for which there are no error locators θ, η with $\mu_A(\theta)=r$ or $\mu_{A'}(\eta)=r'$.

On the other hand if there are error locators θ' and η', with $\mu_A(\theta')=r$ and $\mu_{A'}(\eta')=r'$, then the true value $\theta\chi_{r'} \cdot e = \theta'\chi_{r'} \cdot e = 0$ and similarly the true value $\psi_r\eta \cdot e = \psi_r\eta' \cdot e = 0$. Hence the guess must produce the correct value for $\psi_r\chi_{r'} \cdot e$, and hence the correct vote for $\varphi_s \cdot e$. Thus, the number of correct votes is at least equal to the number p of pairs for which there are error locators of both orders.

Hence the difference between the number of correct votes and the number of incorrect votes is at least $p-q$. By Proposition 7.3 we know that $p-q \geq u-t = s-2\gamma-2t+1$ as claimed. □

COROLLARY *The existence of a basis element φ_s for $s>2t+2\gamma-1$ implies the existence of an entry of order s in the syndrome table. Indeed φ_s can be chosen as a product $\psi_r\chi_{r'}$.* □

EXAMPLE In our example the lowest order of an unknown entry is $s=12$. There is only one test entry, and its estimate is correct. We can use this entry to augment the syndrome table by calculating the order 12 entries from the value of $\varphi_{12} \cdot e$ determined by that entry. The new table is shown below.

	0	3	5	6	7
0	0	5	12	11	9
3	5	11	5	7	14
5	12	5	14	14	11
6	11	7	14	7	
7	9	14	11		

The basic idea of the correction algorithm is to continue to fill in entries in the table using this proposition until an error locator is found. You should notice that the fact that no error locator has been found is an essential assumption in the proofs. The calculations are not valid if there is a known error locator.

7.7 The Duursma error processing algorithm

Following the pattern of Chapter 6 this paragraph provides a formal description of the DU-algorithm. It is followed by a verification and an extensive example.

ALGORITHM Let c be a code word of the code $C_\Omega(B, D)$ based on a curve of genus γ and let $f = c + e$, where e has weight $\leq t$ and $d(D) \geq 2t + 2\gamma - 1$.

Step 0 Preliminary calculations

This step is performed once only for any given code.

Let Q be a rational point such that $B(Q) = 0$ and let $D' = D - (d(D) + 2\gamma + 2t - 1)Q$ and choose a divisor A such that $d(A) = t$. Put $A' = D' - A - (2\gamma - 1)Q$, so that $D' = A + A' + (2\gamma - 1)Q$.

Choose bases $\{\varphi_0, \ldots, \varphi_{3\gamma+2t-1}\}$ of $L(D' + \gamma Q)$, $\{\psi_0, \ldots, \psi_{t+2\gamma-1}\}$ of $L(A + (2\gamma - 1)Q)$, and $\{\chi_0, \ldots, \chi_{t+2\gamma-1}\}$ of $L(A' + (2\gamma - 1)Q)$. The bases are indexed by their μ values and each have γ gaps.

If $D = aQ$ then the bases can all be chosen as subsets of a single basis of $L(D' + \gamma Q)$.

For all indices j, k such that $2\gamma + 2t - 1 < j + k \leq 3\gamma + 2t - 1$, express the product $\psi_j \chi_k$ as a linear combination of the basis φ_i.

Step 1 Syndrome calculation

Given a received word f calculate the syndromes $S_{jk} = \psi_j \chi_k \cdot f$ for all values j, k such that $j + k \leq d(D)$. If all syndromes are 0, then f is a code word. Construct a $(t + \gamma) \times (t + \gamma)$ table, with the rows indexed by the basis ψ_j and the columns indexed by the basis χ_k. The row and column indices go from 0 to $t + 2\gamma - 1$ with γ gaps. Enter the calculated syndrome values in this table, leaving cells of order greater than $d(D)$ blank.

Calculate also the syndromes $\varphi_i \cdot e = \varphi_i \cdot f$ for $i \leq d(D)$.

Step 2 Test for error locator

Suppose that all the entries of order less than s are known, but the entries of order s are unknown. Then the rows and columns of order less than $u = s - 2\gamma - t + 1$ are known in the sense that all their entries are known.

Find a non-zero solution of the linear equations

$$\sum_{j<u} S_{jk} \alpha_j = 0, \qquad k \leq 2\gamma + t - 1.$$

If such a solution α exists let $\theta = \sum \alpha_j \psi_j$. Then θ is an error locator for e with $\mu_A(\theta) < u$.

If no such solution exists find a non-zero solution of the linear equations

$$\sum_{k<u} S_{jk} \beta_k = 0, \qquad j \leq 2\gamma + t - 1.$$

If such a solution β exists let $\eta = \sum \beta_k \chi_k$. Then η is an error locator for e with $\mu_{A'}(\eta) < u$. One of these sets of equations will have a non-zero solution at the latest when $u = \gamma + t + 1$.

If an error locator has been found proceed to Step 5; otherwise continue with Step 3.

Step 3 Estimate additional syndromes
For each pair r, r' with $r + r' = s$, a row estimate is a value $S_{rr'}$ that makes the last row of $S|_{r,r'}$ a linear combination of its predecessors. Such a value exists if and only if the last row of $S|_{r,r'-1}$ is a linear combination of its predecessors, and that is the case if and only if there is a row candidate of order $(r, r' - 1)$. Calculate it by solving

$$\sum_{j<r} S_{jk}\alpha_j = S_{rk}, \quad k < r'.$$

If a non-zero solution α exists, put $S_{rr'} = \sum_{j<r} \alpha_j S_{jr'}$. A column estimate is analogous. It makes the last column of $S_{rr'}$ a linear combination of its predecessors and exists if and only if there is a column candidate of order $(r - 1, r')$. It is found by solving

$$\sum_{k<r'} S_{jk}\beta_k = S_{jr'}, \quad j < r.$$

If a non-zero solution β exists, put $S_{rr'} = \sum_{k<r'} \beta_k S_{rk}$.

If both estimates exist, they are unique and agree and we call $S_{rr'}$ a *test entry*. Otherwise we discard the estimate.

Step 4 Majority voting
For each test entry $S_{rr'}$ found in Step 3, use the expression of $\psi_r\chi_{r'}$ in terms of the basis φ_i and the known syndromes $\varphi_i \cdot e$ for $i < s$ to calculate the vote $\varphi_s \cdot e$.

The correct value $\varphi_s \cdot e$ is the one that occurs most frequently. With this value of $\varphi_s \cdot e$ recalculate all the syndromes $S_{rr'}$ with $r + r' = s$ which gave no estimate or an incorrect value for $\varphi_s \cdot e$ by means of the expressions of $\psi_r\chi_{r'}$ in terms of the basis φ_i. Enter these values in the table.

If additional rows or columns have been completed return to Step 2. Otherwise return to Step 3.

Step 5 Find error locations
We assume that we have found an error locator θ with $\mu_A(\theta) < t + \gamma + 1$. Determine the set $M \subseteq B$ of P_l for which $\theta(P_l) = 0$. This set contains the error locations.

Step 6 Calculate error values
If an error locator has been found using the known entries of order less than s and M is the set found in Step 5, solve the equations

$$\sum_{P_l \in M} \varphi_i(P_l)e_l = \varphi_i \cdot f, \qquad i < s.$$

Extend the unique solution $(e_l : P_l \in M)$ of this set of equations by putting $e_l = 0$ for $P_l \notin M$. Then $e = (e_1, \ldots, e_n)$ is the error word.

7.8 Verification

THEOREM *Let c be a code word of the Goppa code $C_\Omega(B, D)$ based on a curve of genus γ and let $f = c + e$. Suppose that e has weight at most t and that $d(D) \geq 2t + 2\gamma - 1$. Suppose further that Q, D', A and A' have been constructed satisfying the conditions of Step 0 of the DU-algorithm, and that the syndrome table has been set up as in Step 1 of the algorithm.*

(a) *If the first t rows of the syndrome table are known then the first set of equations in Step 2 has a non-zero solution.*

(b) *If the known entries of the table have order less than s and the first equations of Step 2 have a solution x then the function $\theta = \sum \alpha_j \psi_j$ is an error locator in $L(A + (s - t - 2\gamma)Q)$.*

(c) *If the known entries of the table have all possible orders less than s and yield a non-zero solution to the first equations of Step 2, then the entries on the right hand side of the equations of Step 6 are all known, and the error word e is the unique solution of these equations.*

Mutatis mutandis, *statements* (a), (b), *and* (c) *hold for columns.*

Suppose the known entries of the table have all possible orders less than s and yield no non-zero solutions for either the first or the second set of equations in Step 2 and that there are unknown entries of order s in the table.

(d) *If row and column estimates both exist for an entry of order s, they are unique and agree. Such an entry is called a test entry.*

(e) *There are more test entries producing a correct estimate for $\varphi_s \cdot e$ than ones producing an incorrect estimate.*

Proof (a) The first t rows of the table contain the syndromes $\psi_j \chi_k \cdot e$, where the functions ψ_j form a basis of $L(A+rQ)$ which thus has dimension t. Applying Proposition 6.2, it follows that there exists a non-zero error locator in $L(A+rQ)$. If $\theta = \sum \alpha_j \psi_j$ is such an error locator, then the values α_j form a solution of the first set of equations in Step 2.

(b) If the first equations in Step 2 yield a non-zero solution x then the function $\theta = \sum \alpha_j \psi_j$ satisfies $\theta \chi_k \cdot e = 0$ for all χ_k. The functions χ_k form a basis of $L(D' - A)$ so $\theta \chi \cdot e = 0$ for all $\chi \in L(D' - A)$. The degree of $D' - A$ is $t + 2\gamma - 1$ by construction and so by Proposition 6.3, θ is an error locator. The known rows are those with order less than $s - t - 2\gamma + 1$, so the corresponding functions ψ_j form a basis of $L(A + (s - t - 2\gamma)Q)$.

(c) The entries on the right hand side of the equations in Step 6 are the syndromes corresponding to the functions φ_r for $r < s$. For values less than $2t + 2\gamma$ the values are known from the received word. For values greater than $2t + 2\gamma - 1$ the existence of a function φ_r implies the existence of a table entry of order r by Cor .7.6 and all these entries are known by hypothesis. Proposition 7.5 then tells us that the syndromes $\varphi_r \cdot e$ are all known. By Proposition 6.2 e is then the unique solution of the equations.

(d) This follows from Proposition 7.4c (see also Exercises 8 and 9).

(e) This is just a restatement of Proposition 7.6. □

7.9 A complete example

EXAMPLE To demonstrate the algorithm we complete the calculations starting from the extended table we found in Paragraph 7.6.

At the end of Example 7.6 no new rows became fully known, so there is still no error locator and we must estimate the entries of order 13; there are four of these, all yielding the syndrome $\varphi_{13} \cdot e$ directly. By symmetry the calculations for the entries S_{67} and S_{76} are the same and so are those for the entries S_{58} and S_{85}.

We begin with the calculations for S_{85}. There are several row candidates giving different estimates for S_{85}, for instance $\psi_8 + \psi_3 + \psi_0$, which gives the correct value $S_{85} = 9$, and $\psi_8 + 8\psi_5 + 15\psi_3$, which gives the wrong value $S_{85} = 8$. Of course we only know that one of the values is correct because we already know the error word. As expected there is no column candidate. That follows from the fact that there was no column candidate for S_{75}. So we could have saved ourselves these calculations.

For S_{76} we get several row candidates, for instance $\psi_7 + \psi_5 + \psi_3$ and $\psi_7 + 4\psi_6 + 12\psi_3 + 7\psi_0$ and a unique column candidate $\chi_6 + 6\chi_5 + 5\chi_3 + 11\chi_0$. As stated in Proposition 7.4 they all give the same value $S_{76} = 9$. We thus have two test entries both giving the correct guess. Indeed, since at this stage correct guesses must outnumber incorrect ones by two, and the table is symmetric, it follows a priori that all test entries will give correct values.

We now have a syndrome table with a new known row (and by symmetry also a new known column).

	0	3	5	6	7	8
0	0	5	12	11	9	5
3	5	11	5	7	14	14
5	12	5	14	14	11	9
6	11	7	14	7	9	
7	9	14	11	9		
8	5	14	9			

So we test for an error locator. The fact that there was no column candidate for S_{75} (and by symmetry also no row candidate for S_{57}) tells us that the search will be unsuccessful. The known rows are still linearly independent. So we must continue and estimate the entries of order 14.

This time the number of correct test entries must outnumber the incorrect ones by three. Since there are only three entries of order 14 they must all be correct test entries and we need only calculate one of them. However, I shall calculate all three to verify this fact directly. The entries are $S_{86} = S_{68}$ which give $\varphi_{14} \cdot e$ directly and S_{77} which does not.

Begin with S_{86}. Again there are several row candidates, of which one is $\psi_8 + \psi_3 + \psi_0$, and a unique column candidate, which is necessarily the same as the one for S_{76}, namely $\chi_6 + 6\chi_5 + 5\chi_3 + 11\chi_0$. All of these candidates give the value $S_{86} = 12$.

Now take S_{77}. By symmetry any row candidate is also a column candidate. Such a candidate is $\psi_7 + \psi_5 + \psi_3$ which gives the value $S_{77} = 5$. From the table in Paragraph 7.2 we see that this is the syndrome of $\varphi_{14} + \varphi_7$. The syndrome of $\varphi_7 = 9$ is known and so we obtain $\varphi_{14} \cdot e = 9 + 5 = 12$ which agrees with the estimate from S_{86}. This extends the table to

	0	3	5	6	7	8
0	0	5	12	11	9	5
3	5	11	5	7	14	14
5	12	5	14	14	11	9
6	11	7	14	7	9	12
7	9	14	11	9	5	
8	5	14	9	12		

We have an additional known row and column and the row candidate for S_{68} (which was a correct test entry) gives us an error locator $\varphi_6 + 6\varphi_5 + 5\varphi_3 + 11\varphi_0$.

The next step is to find the error locations from this error locator. For that we add the appropriate multiples of the rows of Table 5.5.

11 ×	1	1	1	1	1	1	1	1	1	1	1	1	1	1	1	1
5 ×	0	0	11	10	8	15	3	5	11	10	11	10	4	9	14	2
6 ×	0	0	1	1	2	4	9	14	13	7	12	6	7	12	6	13
1 ×	0	0	10	11	15	3	5	8	10	11	10	11	9	14	2	4
	11	11	11	15	11	8	5	4	8	2	14	4	4	0	0	0.

So the error locations M are, as before, the last three points of B:

$$M = \{(13, 15), (7, 3), (12, 5)\}.$$

The final part of the calculations is similar to those used for the SV-algorithm. We use Table 5.5 to determine the set of equations for the error vector and obtain

$$\begin{bmatrix} 1 & 1 & 1 \\ 9 & 14 & 2 \\ 12 & 6 & 13 \\ 14 & 2 & 4 \\ 5 & 8 & 15 \\ 8 & 15 & 3 \end{bmatrix} \begin{bmatrix} x \\ y \\ z \end{bmatrix} = \begin{bmatrix} 0 \\ 5 \\ 12 \\ 11 \\ 9 \\ 5 \end{bmatrix}.$$

These equations have the unique solution $x = 3$, $y = 2$, and $z = 1$, giving us

$$\begin{aligned} e &= 0\,0\,0\;\; 0\;\; 0\,0\,0\,0\;\; 0\,0\,0\;\; 0\,0\,3\,2\,1; \\ c = f - e &= 8\;0\;3\;11\;15\;3\;1\;9\;15\;5\;0\;14\;1\;1\;1\;1. \end{aligned}$$

Thus we have correctly identified the errors and reconstructed the original code word.

7.10 Exercises

1. Prove Corollary 7.1: Let $\varphi_1, \ldots, \varphi_m$ be functions such that

$$\mu_H(\varphi_1) < \cdots < \mu_H(\varphi_m)$$

is a complete list of the possible values $\mu_A(\psi) \leq r + d(H)$. Then $\varphi_1, \ldots, \varphi_m$ form a basis of $L(H + rQ)$.

2. Prove that if $K = H + nQ$ for any integer n, then $\mu_K(\theta) = \mu_H(\theta)$ for all functions θ. Deduce that if $H = dQ$, then $\mu_H(\theta) = -\nu_Q(\theta)$ for all functions θ.

3. Verify the claims of Example 7.3. That is, check that the functions $\varphi_6 + 11\varphi_0 + 5\varphi_3 + 6\varphi_5$, $\varphi_7 + \varphi_3 + \varphi_5$, and $\varphi_8 + \varphi_0 + \varphi_3$ are error locators for the word $e = 0000000000000321$ and that there are no error locators θ with $\mu_A(\theta) \leq 5$ for $A = 3Q$. The calculations can be performed using the rows of Table 5.5.

4. A Hermitian example. This exercise continues the work of Exercise 15 of Chapter 4 and Exercises 7 and 4 of Chapter 5. Consider the Hermitian curve $x^5 + y^4 + y = 0$ defined over the field F_{16}. Let Q be the unique point at infinity of the curve and let B be the sum of the 64 other rational points. Suppose the code $C_\Omega(B, 21Q)$ is used for transmission and a code word undergoes an error of weight 5 as indicated below, where the first row indicates the point of B where the error occurs and the second the error value at that point.

$(1,6)$	$(1,7)$	$(4,4)$	$(4,5)$	$(2,2)$
5	4	3	2	1

Using the auxiliary divisor $A = 5Q$ and Q as the distinguished rational point, calculate the complete syndrome table for this error, filling in all values up to order 27.

5. A Hermitian example (cont.). If the error of the previous question had been added to an arbitrary code word, only the entries in the syndrome table of order 21 or less would be known. Using these entries as a starting point apply the DU-algorithm and verify that it returns the correct error word.

6. Consider the same code as in Example 7.3 and assume that the error word is $e = 1010000000000100$. With the divisor $A = 3Q$ and the point Q of the example find the values $s \leq 8$ for which there exist error locators θ with $\mu_A(\theta) = s$. Draw up a table similar to the one in Paragraph 7.3, listing for the values u between 3 and the smallest value s the pairs r, r' with $u \leq r \leq 8$ and $r + r' = 8 + u$.

Calculate for each u the number p of such pairs for which there exist error locators θ, θ' with $\mu_A(\theta) = r$ and $\mu_A(\theta') = r'$ and also the number of pairs q for which neither an error locator θ with $\mu_A(\theta) = r$ nor one with $\mu_A(\theta) = r'$ exists. Verify that $p - q \geq u - 3$ in all cases.

7. The DU-algorithm is to be used to correct up to t errors for a one-point code $C_\Omega(B, D)$ where $D = aQ$. Show that if Q is used as the auxiliary point, D' is constructed as in Paragraph 7.1, and A is chosen to be tQ, then $A' = A$.

8. Suppose that the syndrome table entries of order less than s are known and that $r + r' = s$. Let $\theta = \psi_r + \theta'$ be a normalized row candidate of order (r, r') and $\eta = \chi_{r'} + \eta'$ be a normalized column candidate of order (r, r'). Show that

$$\psi_r \chi_{r'} \cdot e = -\theta' \chi_{r'} \cdot e = -\psi_r \eta' \cdot e = \theta\eta \cdot e + \theta'\eta' \cdot e.$$

Explain why the known entries of the table give the same value for $\theta' \chi_{r'} \cdot e$ and $\psi_r \eta' \cdot e$.

9. Suppose that all the entries of an $m \times n$ matrix $A = (a_{ij})$ are known except a_{mn} and let A' be the $(m-1) \times n$ matrix consisting of the first $m-1$ rows of A. Suppose that there are two different values for a_{mn} that make the last row of A a linear combination of its predecessors. Show that the last column of A' is linearly independent of its predecessors and that there is thus no value for a_{mn} that makes the last column of A a linear combination of its predecessors.

Show further that if there is a value $a_{mn} = a$ that makes the last row of A a linear combination of its predecessors and a value $a_{mn} = b$ that makes the last column of A a linear combination of its predecessors, then $a = b$.

Explain the relation between this result, Prop 7.4, and Exercise 8.

Part II

Fields of algebraic functions

8

INTRODUCTION: THE ALGEBRAIC APPROACH

In Part I we saw how the algebraic structures associated with a curve can be used to construct good codes and error processors for them. The main aim of this part is to prove the theorems stated in Chapter 4. The simplest proofs by far are those that look only at the algebra of the function field of the curve.

8.1 Function fields

To that end it is necessary to define a function field by its internal properties rather than by reference to a curve. Then one must see what corresponds to a point (or rather a conjugacy class of points) in the function field of the curve. That abstract concept will be based on finding an equivalent for the local ring of a point, called a *place*. Every place determines an order function so the problems of singularity disappear. A singular point corresponds to a whole set of places rather than a single one.

In this guise the theory bears a close resemblance to complex function theory and I shall draw on that theory also to provide motivation and examples. Readers not familiar with complex function theory can simply ignore these references. For them the motivation may work in the opposite direction if they study that theory in the future. The idea of replacing the geometric theory of curves by the algebraic theory of function fields was pioneered in the masterly treatise of Chevalley (1951) and taken up and simplified by Roquette (1958) and his school (Deuring 1973, Stichtenoth 1978, 1991). The present book introduces some further simplifications.

8.2 Pros and cons

An additional advantage of this abstract approach is that it covers curves in n-dimensional space as well as plane curves. This is important, because the number of rational points on a plane curve over K is bounded by $|K|^2$. That makes it impossible to construct arbitrarily long codes over plane curves. The restriction is much less stringent then for Reed–Solomon codes where the block length is bounded by $|K|$ but it does make it impossible for Goppa codes from plane curves to meet the Gilbert bound.

On the other hand, a disadvantage of the approach is that to understand it one needs to be comfortable with the structures of algebra as taught in a first course on higher algebra. In particular, the reader should be familiar with ideals of commutative rings and the basic theory of fields. No great depth of knowledge is required. Any standard undergraduate text in algebra will contain everything required. The classic text by Birkhoff and Maclane (1977) covers more than enough material. For a more modern text I would recommend the text by Cohn, particularly its first volume (1982).

This part also assumes a greater level of mathematical sophistication than the first part. The proofs are less expansive and the examples are more general.

8.3 Chapter layout

Chapter 9 lays the basis by providing the fundamental definitions. After that Chapter 10 develops the theory of discrete valuations, which correspond to order functions for points. It contains a proof of the Artin–Whaples approximation theorem, an important tool in the later chapters. Chapter 11 introduces divisors and L-spaces in direct analogy to Chapters 3 and 4. It contains proofs of the degree theorem and Riemann's theorem. Chapter 12 deals with repartitions and differentials. Although I have somewhat simplified the presentation of this chapter in comparison with the standard literature it remains the hardest of the book. It contains proofs of the full Riemann–Roch theorem and the residue theorem and of the fact that Goppa function codes and Goppa residue codes are the same. It is followed by a chapter (Chapter 13) on extending fields. This covers three topics: the first is quite elementary and discusses the effect of changing the base field without changing the field of functions. The next topic is of importance in the theory of codes and considers what happens if the function field is extended only by increasing the set of constants. That occurs if a curve defined over a small field is analysed over a larger field. The last part of the chapter takes the first steps in the analysis of true extensions of function fields.

In Chapter 14 we return to curves, proving the horizon theorem and the Plücker formula for the genus. The horizon theorem requires a curve in order to define the meaning of the horizon. The last chapter discusses the Gilbert bound and the existence of curves yielding codes meeting or coming close to that bound.

The Plücker formula does not have a direct equivalent for curves in higher dimensions. A major problem for the application of curves to coding theory is the absence of any easy way to determine the genus of a function field or general curve. The general theorems giving estimates require a deepening of the topics discussed in the last part of Chapter 13. The reader who wishes to learn more about this topic is referred to the books of Stichtenoth (1991) and Tsfasman and Vlăduţ (1991). These general theorems are more difficult than anything in this book but they are still not sufficiently powerful to determine the exact genera of the curves discussed in the last chapter.

9

FUNCTION FIELDS AND PLACES

This chapter introduces the abstract algebraic concepts that correspond to the function fields of curves and the local rings of points on the curves. The field of rational functions, which corresponds to a straight line, is examined in detail. The chapter concludes with the existence theorem which shows that a function field contains enough structure to serve as an algebraic equivalent for a curve.

9.1 Function fields of plane curves

How can one describe the function field of a curve without referring to the curve? In other words, what intrinsic properties characterize function fields of curves? Consider an example.

EXAMPLE $C\colon x^3 + y^3 + 1 = 0$.

If we regard x as the 'basic variable' then y is a root of the polynomial $x^3 + y^3 + 1 = 0$ defined over $K(x)$. On the other hand, x is not a root of a polynomial with coefficients in K. So $K(x)$ is just the field of rational functions over K: $K(x) \cong K(\dot{x})$. As $x^3 + y^3 + 1$ is irreducible over K, it is irreducible over $K(x)$. Thus $K(C)$ is an extension of $K(x)$ of degree 3.

It is clear that this procedure works for any curve whose equation involves y. All irreducible polynomials must involve x or y, so for irreducible polynomials not involving y we can just reverse the roles of x and y. It is an easy standard exercise to show that a finite extension is algebraic, and that the converse is not true. We thus have the following proposition.

PROPOSITION *Let C be a plane algebraic curve defined over the field K. Then the function field $K(C)$ contains an element x that is not algebraic over K such that $K(C)$ is a finite extension of $K(x)$.* □

This property will form the basis of our abstract definition. For reference it is useful to recall here the definition of algebraic and transcendental elements and the dot convention.

DEFINITION We denote by $K[\dot{x}]$, $K[\dot{x}, \dot{y}]$ the formal polynomial rings over the field K. For an element x in and extension field E of K, $f(x)$ denotes the evaluation of the polynomial $f(\dot{x})$ at x. The element is called *algebraic* if it is a root of a polynomial in $K[\dot{x}]$. Otherwise it is called *transcendental.*

The field $E \supseteq K$ is called an *algebraic extension* if all elements of E are algebraic over K. The *degree* $|E : K|$ is the vector space dimension of E over K. The extension is called *finite* if $|E : K|$ is finite.

9.2 Algebraic functions

Here then is the formal definition of a function field.

DEFINITION Let K be a field. An extension field F of K is called a *field of algebraic functions of one variable* or *function field* if F contains an element x transcendental over K such that $|F : K(x)|$ is finite. In the special case that $F = K(x)$ it is called a *field of rational functions*, because it is isomorphic to the field of rational functions $F(\dot{x})$.

The field $\bar{K}$ of elements in F that are algebraic over K is called the field of *constants*.

Notice that $|\bar{K} : K| = |\bar{K}(x) : K(x)| \leq |F : K(x)|$. So $\bar{K}$ is a finite extension of K. We denote the degree $|\bar{K} : K|$ of this extension by κ and call it the *constant degree* of F over K.

Many texts assume that $K = \bar{K}$, which is true for function fields of algebraic curves over finite fields, but may not hold for infinite K. It is even common to assume that K is algebraically closed, which is never true for finite fields K. In fact, all results carry over to the general case where $\bar{K} \neq K$ with only the occasional appearance of an explicit κ. So we shall not make any general assumptions about K. From now on, we shall deal with the theory that flows from this definition, and return to the specific case of function fields of curves in Chapter 14.

The element x is not really special in F. Its part can be taken by any element y outside $\bar{K}$.

PROPOSITION *Let F be an algebraic function field over K with field of constants $\bar{K}$. Then $|F : K(y)|$ is finite for all $y \in F \setminus \bar{K}$.*

Proof Since $|F : K(x)| < \infty$, the powers of y are not linearly independent over $K(x)$. So y is the root of a polynomial f with coefficients in $K(x)$. Multiplying by a common denominator, we can assume that the coefficients of f are all polynomials in x. Hence f can be rewritten as a polynomial $f(\dot{x}, \dot{y})$ in two indeterminates with coefficients in K such that $f(x, y) = 0$. If $\dot{x}$ does not occur in $f(\dot{x}, \dot{y})$, then y is algebraic over K and thus lies in $\bar{K}$ against our assumption. So $\dot{x}$ appears in $f(\dot{x}, \dot{y})$ and it follows that x is algebraic over $K(y)$. Thus $|K(x, y) : K(y)| \leq \deg_{K(y)} f(\dot{x}, y)$, which is finite. On the other hand $|F : K(x, y)| \leq |F : K(x)|$. Therefore $|F : K(y)| = |F : K(x, y)|\,|K(x, y) : K(y)|$ and this product is finite. □

DEFINITION For any non-constant y we denote by $K[y]$ the set of polynomials in y with coefficients in K and by $K(y)$ its quotient field, the set of rational functions in y. Observe that as y is not algebraic over K, $K[y]$ is isomorphic to the 'abstract' polynomial ring $K[\dot{y}]$.

9.3 The field defined by an irreducible polynomial

In this paragraph we give a construction that gives rise to most function fields over K. The construction is based on an irreducible polynomial $f(\dot{x}, \dot{y})$ which corresponds to the polynomial defining a plane curve. The following proposition is needed to legitimize the construction.

PROPOSITION *Suppose that $K[\dot{x}, \dot{y}]$ is a formal ring of polynomials in two indeterminates and let $f(\dot{x}, \dot{y})$ be an irreducible polynomial in $K[\dot{x}, \dot{y}]$. Then the residue class ring $K[x, y]$ of $K[\dot{x}, \dot{y}]$ modulo $f(\dot{x}, \dot{y})$ is an integral domain generated by x and y. Provided that $f(\dot{x}, \dot{y}) \notin K[\dot{x}]$, the subring $K[x]$ of $K[x, y]$ is isomorphic to $K[\dot{x}]$.*

Proof That the residue class ring is generated by x and y is immediate from the fact that every element of $K[\dot{x}, \dot{y}]$ is a polynomial in $\dot{x}$ and $\dot{y}$. That it is a domain follows from Paragraph 2.3, where it is shown that if an irreducible polynomial divides a product it divides one of the factors. So if $g(x, y)h(x, y) = 0$, then $g(\dot{x}, \dot{y})h(\dot{x}, \dot{y})$ is a multiple of $f(\dot{x}, \dot{y})$. Since $f(\dot{x}, \dot{y})$ is assumed to be irreducible, it follows that one of $g(\dot{x}, \dot{y})$ and $h(\dot{x}, \dot{y})$, say $g(\dot{x}, \dot{y})$, is a multiple of $f(\dot{x}, \dot{y})$. That implies that $g(x, y) = 0$.

It remains to show that no polynomial in x alone is zero. But if $g(x) = 0$, then $g(\dot{x})$ is a multiple of $f(\dot{x}, \dot{y})$. Considering $K[\dot{x}, \dot{y}]$ as a polynomial ring in $\dot{y}$ over $K[\dot{x}]$, it follows that $\deg_{\dot{y}} g(\dot{x}) \geq \deg_{\dot{y}} f(\dot{x}, \dot{y})$. By hypothesis, $f(\dot{x}, \dot{y}) \notin K[\dot{x}]$, so $\deg_{\dot{y}} f(\dot{x}, \dot{y}) > 0$. On the other hand $g(\dot{x}) \in K[\dot{x}]$, and the only element of this ring with $\dot{y}$-degree greater than 0 is 0. So $g(\dot{x}) = 0$. □

COROLLARY *The field of fractions $K(x, y)$ of $K[x, y]$ is a field of algebraic functions.*

Proof As $f(\dot{x}, \dot{y})$ is irreducible, it does not lie in $K = K[\dot{x}] \cap K[\dot{y}]$. Hence it does not lie in one of $K[\dot{x}]$ and $K[\dot{y}]$, say $K[\dot{x}]$. Then by the proposition $K[x]$ is isomorphic to $K[\dot{x}]$ and y is a root of $f(x, \dot{y})$. Hence x is transcendental over K and $|K(x, y) : K(x)| \leq \deg_{\dot{y}} f(\dot{x}, \dot{y})$. □

In fact the degrees are equal, because $f(x, \dot{y})$ is irreducible over $K[x]$, and that implies irreducibility over $K(x)$ by Gauss' lemma (2.2).

This method of construction gives us a very large class of function fields. Indeed, if F is a function field that is *separable* over $K(x)$, then by the theorem of the primitive element it has the form $K(x, y)$ for a single element y and can therefore be constructed as the field of fractions of a homomorphic image of $K[\dot{x}, \dot{y}]$. Separable field extensions are defined in Chapter 13, but the theorem of the primitive element is not proved in this book. It forms part of a standard course on Galois theory and can be found in the books of Birkhoff and MacLane (1977) and Cohn (1977).

Such function fields are called *separably generated.* It is possible to show (Stichtenoth 1991) that if K is perfect, then every function field over K is separably generated. A field is *perfect* if every finite extension of K itself is separable. Perfect fields include all finite fields and all fields of characteristic 0 such as the rationals Q and the reals R.

9.4 Places of a function field F

In this paragraph we shall define a place of a function field. For smooth curves places correspond to conjugacy classes of points, but at singular points there may be several places. Geometrically one can imagine this as slightly separating the different branches of the curve where it crosses itself.

The definition of a place is based on the local ring $K[C]_P$ of a simple point P given in Chapter 3. The ring is defined in Paragraph 3.6 and in Paragraph 3.9 a point is defined to be simple if one of f and $1/f$ lies in $K[C]_P$ for any function f. It is this property that we shall use in the abstract setting.

DEFINITION A subring $V \subseteq F$ is called a *place ring* of F if the following conditions are satisfied:

PL1. $K \subset V \subset F$ (proper inclusion); and

PL2. For all $x \in F$, one (at least) of $x \in V$ and $x^{-1} \in V$ holds.

EXAMPLE The local ring of a point.

Consider the function ring of the curve C: $x^3 + y^3 = 1$ defined over F_2 and the point P: $(10, 0)$ with coefficients in F_4. The appropriate order function for this point is the y-order, because

$$(x - 10)/y = (x - 10)y^2/y^3 = (x - 10)y^2/(x^3 + 1) = y^2/(x^2 + 10x + 11)$$

and the denominator $x^2 + 10x + 11$ is non-zero at P (see Chapter 3). We proved in Paragraph 3.11 that this implies that the local ring $\mathsf{F}_4[C]_P$ satisfies the conditions for a place, but what about the local ring over F_2? This requires no further calculation, because, by its very definition, $\mathsf{F}_2[C]_P = \mathsf{F}_4[C]_P \cap \mathsf{F}_2[C]$. So if $\mathsf{F}_4[C]_P$ is a place of $\mathsf{F}_4[C]$, then automatically $\mathsf{F}_2[C]_P$ is a place of $\mathsf{F}_2[C]$ (see Exercise 2).

The point Q: $(11, 0)$ also lies on the curve. Its local ring over F_4 is different from $\mathsf{F}_4[C]_P$ because it contains $1/(x - 10)$. But the two rings intersect in the same subring of $\mathsf{F}_2[C]$. For if $g(x, y)/h(x, y) \in \mathsf{F}_2[C]_P$ we can assume that $h(P) \neq 0$ and $h(x, y) \in \mathsf{F}_2[x, y]$. We write

$$h(x, y) = h_1(x) + yh_2(x, y).$$

Then $h(P) \neq 0$ implies that $h_1(10) \neq 0$. That is equivalent to the statement that the minimal polynomial of 10 over F_2 does not divide $h_1(x)$. That minimal polynomial is $x^2 + x + 1$ which is also the minimal polynomial of 11 over F_2. Thus $h_1(11) \neq 0$ and so $h(Q) \neq 0$. Therefore $g(x, y)/h(x, y)$ lies in $\mathsf{F}_2(C)_Q$. By symmetry the converse also holds and so the two rings are identical. The reason for this is that P and Q are conjugate over F_2, and by their definition, conjugate points have the same local rings.

9.5 Places of the field of rational functions

We shall determine all places of the field $K(x)$ of rational functions over K. This is an important example because, as you will see later in the chapter, every place of a function field F over K is an extension of a place of $K(x)$, for any non-constant element x. In Chapter 10 we shall find as a consequence of the Artin–Whaples approximation theorem that there are only finitely many places of F lying over a given place of $K(x)$ and we can then apply the methods of Chapter 13 to determine them.

EXAMPLE Rational functions over K.
For each irreducible polynomial $p(x) \in K[x]$ we define a set V_p as follows. The rational function $f(x)/g(x) \in V_p$ if and only if after cancellation of common factors $p(x)$ does not divide $g(x)$. We define one further set V_∞ which contains those rational functions $f(x)/g(x)$ with $\deg f(x) \leq \deg g(x)$ (if one representative f/g of φ satisfies this condition, then so does every other representation f'/g').

PROPOSITION *Let $F = K(x)$ be the field of rational functions over the field K. The place rings of F are precisely the sets V_p for irreducible polynomials $p(x)$ and the set V_∞. These are all distinct.*

Proof (1) We first verify that these sets are indeed places. From the definition it is clear that they satisfy PL1. To check PL2 for V_p, we observe that if $f(x)$ and $g(x)$ are relatively prime, then $p(x)$ cannot divide both of them. So at least one of $f(x)/g(x)$ and $g(x)/f(x)$ lies in V_p. For V_∞ axiom PL2 is almost trivial. Since one of $\deg f(x) \leq \deg g(x)$ and $\deg g(x) \leq \deg f(x)$ must hold.

To complete the verification we must check that V_p and V_∞ are rings. For that we must show that they are closed under addition and multiplication and contain K. The element of $a \in K$ can be written as $a/1$ and this representation satisfies the conditions for both V_p and V_∞. So the sets contain K. The fact that they are closed under addition and multiplication follows from the formulae

$$f_1/g_1 \times f_2/g_2 = f_1f_2/g_1g_2 \quad \text{and} \quad f_1/g_1 + f_2/g_2 = (f_1g_2 + f_2g_1)/g_1g_2.$$

For V_p observe that $p(x)$, being irreducible, does not divide g_1g_2 if it does not divide g_1 or g_2. For V_∞ closure follows because the assumptions $\deg f_1 \leq \deg g_1$ and $\deg f_2 \leq \deg g_2$ imply

$$\deg f_1f_2 \leq \max\{\deg f_1g_2, \deg f_2g_1\} \leq \deg g_1g_2.$$

(2) Now let V be an arbitrary place ring of $F = K(x)$. Consider first the case that $x \in V$. It follows that $K[x] \subseteq V$. It is not possible that $1/f(x) \in V$ for all non-zero polynomials $f(x)$ for then $f(x)/g(x) \in V$ would hold for all non-zero rational functions as well, violating PL1. So choose a non-constant polynomial $p(x) \in K[x]$ of minimal degree such that $1/p(x) \notin V$. If $p = fg$ and both f and g are non-constant, then $\deg f < \deg p$ and $\deg g < \deg p$. So by the choice of p, both $1/f(x) \in V$ and $1/g(x) \in V$. That implies $1/fg = 1/p \in V$, a contradiction. Hence p is not the product of two non-constant polynomials and so it is irreducible.

(3) We shall show that $V_p = V$. First we show that for a polynomial $1/f(x) \notin V$ holds if and only if p divides f. If $f(x) = p(x)q(x)$, then $1/p = q(1/f)$. As $K[x] \subseteq V$ it follows that $1/f \in V$ would imply $1/p \in V$. On the other hand if p

does not divide f, then the highest common factor of f and p is 1 and we can write

$$1 = u(x)f(x) + v(x)p(x)$$

for certain polynomials $u(x)$, $v(x)$. Then $uf/p = (1/p) - v$ and since $v \in K[x] \subseteq V$ while $1/p \notin V$, it follows that $uf/p \notin V$. Hence $p/uf \in V$ and so $p/f = u(p/uf) \in V$. Now

$$1/f(x) = u(x) + v(x)(p(x)/f(x))$$

represents $1/f$ as a sum of two elements in V.

That establishes that V contains all rational functions of the form f/g where p does not divide g. On the other hand, if p divides g and not f then V cannot contain f/g, because then it would contain $1/g = (1/f)(f/g)$. Thus $V = V_p$.

(4) We turn now to the case that $x \notin V$. It follows that $z = 1/x$ is in V. Hence all of the polynomials $K[z]$ lie in V. We can now repeat the argument of the previous case with z as our choice of polynomial p, because $x = 1/z \notin V$. So $V = \{f(z)/g(z)\}$, where z does not divide $g(z)$.

It remains to show that $V = V_\infty$. So let $f(x)/g(z) \in V$ and suppose that

$$f(z) = a_0 + a_1 z + \cdots + a_m z^m,$$
$$g(z) = b_0 + b_1 z + \cdots + b_n z^n.$$

If $k \geq m$ and $k \geq n$ and we multiply both $f(z)$ and $g(z)$ by x^k we will obtain polynomials in x. The degree of $g(z)x^k$ as a polynomial in x is k because $b_0 \neq 0$. The degree of $f(z)x^k$ is $k - i$, where i is the least index such that $a_i \neq 0$. Thus $V \subseteq V_\infty$.

Conversely, let $f(x)/g(x) \in V_\infty$. Multiplying both $f(x)$ and $g(x)$ by z^k, where $k = \deg g(x)$, will produce polynomials in z (since $\deg g(x) \geq \deg f(x)$). If furthermore cx^k is the highest term of $g(x)$, then c is the constant term of $g(x)z^k$ and so z does not divide $g(x)z^k$. Thus $f(x)/g(x) \in V$ establishing that $V_\infty \subseteq V$. □

9.6 Structure of places

A place ring V is a *local* ring, that is it has a unique maximal ideal P. The complement of P in V, which we denote by U, then consists of all invertible elements of V. Any one of the three sets U, P, and V determines the other two. So we can use any of the three to describe the place.

PROPOSITION (a) *A place ring V has a unique maximal ideal P which consists of those elements $x \in V$ for which $x^{-1} \notin V$.*
(b) *Either of the sets P or $U = V \setminus P$ determines the ring V.*

Proof (a) We define U to be the set of invertible elements $x \in V$ (that is, $x^{-1} \in V$) and P to be $V \setminus U$. Then trivially two of these sets determine the third. Notice that U is closed under multiplication.

It will be sufficient to show that the non-invertible elements form an ideal. For if an ideal contains an invertible element x it also contains $x^{-1}x = 1$ and so it is the whole ring V. Hence all proper ideals are subsets of the set of non-invertible elements.

So suppose $x \in P$ and $y \in V$. We wish to show that $xy \in P$. If not, then $uxy = 1$ for some $u \in V$. But then $x^{-1} = uy \in V$ contradicting the assumption that $x \in P$. Therefore $xy \in P$.

Suppose that $x, y \in P$. We wish to show that $x + y \in P$. We may assume that $x, y \neq 0$. Then one of x/y and y/x lies in V, say x/y. It follows that $x + y = y(x/y + 1) \in P$.

(b) We already know that two of the sets V, P, and U determine the third, and the definition shows that V determines P and U. We shall show that P by itself determines U and vice versa, so that each of the three sets determines the other two.

The set U consists of all elements x such that neither x nor x^{-1} lies in P. Thus P determines U.

Conversely, U also determines P. Indeed, P consists of all sums $u + v \notin U$ with $u, v \in U$. For these sums lie in V, and so the assumption that they are not in U implies that they lie in P. On the other hand, if $x \in P$ then $1 + x \in V$, but $1 + x \notin P$, because that would imply $1 = (1 + x) - x \in P$. Thus x can be written as a sum of elements in U in the form $(1 + x) + (-1)$.

□

DEFINITION Notation and terminology for places.

We shall call the structures associated with a place the *place ring* in the case of V, the *place ideal* in the case of P, and the *place units* in the case of U. A place is often identified with its place ideal.

The notation we shall use henceforth is as follows. For a place P of the field F, the place ideal is also denoted by P and the place ring is denoted by F_P.

9.7 The existence of places

Every function field contains many places. In fact, if we choose any non-constant $x \in F$ and any place P of $K(x)$ we can 'extend' P to a place of F. The extension is not in general unique, but different places of $K(x)$ cannot extend to the same place of F. That assures us that F has at least as many places as $K(x)$. The following proposition establishes this important property in a slightly more general form.

THEOREM Existence theorem *Let A be a ring such that $K \subseteq A \subseteq F$ and let $0 \neq I \neq A$ be an ideal of A. Then there exists a place P of F such that $A \subseteq F_P$ and $I \subseteq P$. If I is a prime ideal, then P can be found so that $P \cap A = I$.*

The proof relies on a fundamental result of set theory, Zorn's lemma, to establish that a certain family of rings has a maximal member. It then shows that such a maximal ring is the desired place. I state Zorn's lemma here. I hope that you will find its claim plausible. Indeed it is often taken as an axiom of set theory, but it is beyond the scope of this book to discuss the subtle issues of abstract logic that its analysis involves.

ZORN'S LEMMA *Let $\mathcal{S}$ be a family of sets such that for any ascending chain*

$$S_1 \subseteq S_2 \subseteq \ldots$$

of members of $\mathcal{S}$, the union $\bigcup S_i$ also lies in $\mathcal{S}$. Then $\mathcal{S}$ has at least one maximal member T, in other words a set $T \in \mathcal{S}$ such that $T \subseteq U \in \mathcal{S}$ implies that $U = T$. □

Proof (1) Let $\mathcal{V}$ be the family of rings $B \supseteq A$ such that $IB \neq B$, where IB denotes the ideal product (the set of all finite sums of products xy with $x \in I$ and $y \in B$). The family $\mathcal{V}$ is not empty because $A \in \mathcal{V}$. Furthermore for an ascending chain of rings

$$B_1 \subseteq B_2 \subseteq \ldots$$

all lying in $\mathcal{V}$, the union $C = \bigcup B_i \in \mathcal{V}$. Indeed, it is straightforward to check that C is closed under addition and multiplication. Further, if $IC = C$, then $1 \in IC$ and so $1 = \sum x_i y_i$ with $x_i \in I$ and $y_i \in C$. Since the sum is finite, we can choose n so that $y_i \in B_n$ for all i. Hence $1 \in IB_n$, which implies that $IB_n = B_n$. That contradicts the assumption that $B_n \in \mathcal{V}$. Thus the assumption that $IC = C$ is untenable.

By Zorn's lemma it follows that there exists a maximal ring V in $\mathcal{V}$. We shall establish that V is a place ring.

(2) To do that we must show that for any element x of F, at least one of x and x^{-1} must lie in V. Suppose that $x \notin V$; then $V[x] \notin \mathcal{V}$ because V is maximal in $\mathcal{V}$ and so by definition we must have $1 \in IV[x]$. Then we can write

$$1 = \sum_{i=0}^{n} y_i x^i$$

with $y_i \in IV$. If also $x^{-1} \notin V$, then

$$1 = \sum_{j=0}^{m} z_j x^{-j}$$

with $z_i \in IV$. We choose m and n minimal so that these equations hold. Then certainly $m, n \geq 1$ and we may assume by symmetry that $m \leq n$. We multiply the first equation by $1 - z_0$ and the second by $y_n x^n$ and add:

$$1 - z_0 + y_n x^n = \sum_{i=0}^{n} (1 - z_0) y_i x^i + \sum_{j=0}^{m} z_j y_n x^{n-j}.$$

Gathering terms of like degree in x we obtain

$$\begin{aligned}1 = z_0 - (1 - z_0)y_0 \\ + \sum_{i=1}^{n-m-1} (1 - z_0)y_i x^i + \sum_{i=n-m}^{n-1} \big((1 - z_0)y_i + z_{n-i}y_n\big)x^i \\ + \big((1 - z_0)y_n + z_0 y_n - y_n\big)x^n.\end{aligned}$$

The term of degree n is zero. It is also possible that one or other of the sums is empty, or that the second sum contributes an additional term of degree 0, but in all cases this equation has the form

$$1 = \sum_{i=0}^{n-1} w_i x^i$$

with $w_i \in IV$, contradicting the minimality of n. Thus our assumption that both $x \notin V$ and $x^{-1} \notin V$ fails. Hence V is a place ring F_P. Since IV is a proper ideal of V it is a subset of the place ideal P, and so $I \subseteq IV \cap A \subseteq P \cap A$ establishing the first claim of the theorem.

(3) Suppose now that I is a prime ideal. By definition, that means that for elements $x, y \in A \setminus I$, the product $xy \in A \setminus I$. We first extend A and define B to be the ring $\{x/y : x \in A,\ y \in A \setminus I\}$. Then $IB = \{x/y : x \in I,\ y \in A \setminus I\}$ is an ideal of B. Indeed since every element outside IB is invertible in B, IB is the unique maximal ideal of B (so B is a local ring). By its construction $I \subseteq IB \cap A$ and $(A \setminus I) \cap IB = \emptyset$. Hence $IB \cap A = I$.

Applying the first part of the theorem to B we find a place P with $IB \subseteq P$. Now $P \cap B$ is a proper ideal of B as it does not contain 1, so $P \cap B \subseteq IB$. Hence $P \cap B = IB$ and therefore $P \cap A = IB \cap A = I$. □

COROLLARIES (a) *Every non-constant element $x \in F$ is contained in the place ideal of some place P.*
(b) *The set of constants $\bar{K}$ is the intersection of all place rings.*
(c) *There are infinitely many places in F.*

Proofs (a) Apply the theorem to $A = K[x]$ and the prime ideal $\langle x \rangle$.

(b) From (a) it follows that the intersection is contained in $\bar{K}$, because for a non-constant x there is a place ideal P containing x^{-1}. The corresponding place ring does not contain x.

On the other hand, let $x \in \bar{K}$. Then x is algebraic over K, so let $x^n + \sum a_i x^{n-i} = 0$ with $a_i \in K$. If $x \notin F_P$ for some place P, then $x^{-1} \in P$. Therefore

$$1 = x^{-n}x^n = -x^{-n}\left(\sum a_i x^{n-i}\right) = -\sum a_i x^{-i} \in P.$$

That is impossible. Hence $x \in F_P$ for any place P.

(c) $K[x]$ has infinitely many irreducible polynomials p and each generates a different prime ideal $\langle p\rangle$. To each of these ideals there is a place P such that $P \cap K[x] = \langle p\rangle$. □

9.8 Exercises

1. Show that if E/K is a finite extension of degree n, then every element of E is the root of a polynomial of degree $\leq n$ in $K[\dot{x}]$.
2. Consider the function ring of the curve C: $x^3 + y^3 = 1$ defined over F_2 and the point P: $(10, 0)$ with coefficients in F_4. Given that the local ring $\mathsf{F}_4[C]_P$ of P over F_4 is a place of $\mathsf{F}_4[C]$, verify that $\mathsf{F}_2[C]_P = \mathsf{F}_4[C]_P \cap \mathsf{F}_2[C]$ satisfies both conditions for a place of $\mathsf{F}_2[C]$ over $K = \mathsf{F}_2$.
3. Consider the function field $\mathsf{Q}(x, y : x^3 + x^2 - y^2)$, which is the function field $\mathsf{Q}(C)$ of the cubic C defined by $y^2 = x^3 + x^2$. Let P be the origin $(0, 0)$ and let A be the local ring $\mathsf{Q}[C]_P$, which consists of all rational functions that can be written in the form $g(x, y)/h(x, y)$ with $h(0, 0) \neq 0$. Show that neither x/y nor y/x lie in A so that A is not a place (see Chapter 3).

Show that there are exactly two places containing A, one containing x/y and the other containing y/x.
4. Consider the function field $\mathsf{Q}(x, y : x^3 - y^2)$, which is the function field $\mathsf{Q}[C]$ of the cubic C defined by $y^2 = x^3$. Let P be the origin $(0, 0)$ and let A be the local ring $\mathsf{Q}[C]_P$. Show that neither x/y nor y/x lies in A so that A is not a place (see Chapter 3).

Show that there is no place containing A and x/y, and exactly one place containing A and y/x.
5. Let $\mathcal{V}$ be the family of rings defined in the proof of the existence theorem and let V be a maximal ring in $\mathcal{V}$. Prove directly that if $y \in IV$, then $1 + y$ is an invertible element of V. Deduce that IV is the unique maximal ideal of V.

10

VALUATIONS

Having found a satisfactory abstract equivalent for the points P of a curve the next step is to define an equivalent for the order function ν_P. Functions corresponding to good order functions of points are called 'discrete valuations'.

10.1 Discrete valuations

DEFINITION Let F be a field. We define a *discrete valuation* on F to be a map ν from the non-zero elements of F to the integers Z such that

DV1. $\nu(xy) = \nu(x) + \nu(y)$;
DV2. $\nu(x + y) \geq \min\{\nu(x), \nu(y)\}$; and
DV3. $\nu(x) = 1$ for some x.

We extend ν by defining $\nu(0) = \infty$ so that DV1 and DV2 still hold if one of x and y is 0. If $\nu(x) = 0$ for all the non-zero elements of a subfield K of F we call ν a *K-valuation* of F.

PROPOSITION *For any discrete valuation ν we have $\nu(1) = 0$ and for any x, y such that $\nu(x) < \nu(y)$ it follows that $\nu(x + y) = \nu(x)$.*

Proof Let $\nu(t) = 1$. Then

$$1 = \nu(t) = \nu(1t) = \nu(1) + \nu(t) = \nu(1) + 1.$$

From DV2 we have $\nu(x + y) \geq \nu(x)$. But since $x = (x + y) - y$ we also have

$$\nu(x) \geq \min\{\nu(x + y), \nu(-y)\}.$$

Since $\nu(1) = 0$ and $(-1)^2 = 1$ it follows from DV1 that $\nu(-1) = 0$. Hence $\nu(-y) = \nu(y) > \nu(x)$. So $\nu(x) \geq \nu(x + y)$. □

10.2 Valuations and point orders

The x- and y-orders of a point discussed in Chapter 3 are possible discrete valuations. In that chapter we called such an order 'good' if it satisfied an additional condition and investigated the relation between goodness of order functions and simplicity of a point. Here we shall show that an order function is good precisely if it is a discrete valuation. Recall the definitions:

DEFINITION Let C: $f(x, y) = 0$ be a curve and let P: $(x{=}\alpha, y{=}\beta)$ be a point of C with $\alpha, \beta \in K$. Let $\varphi(x, y) \in K(C)$. If $\varphi = (x - \alpha)^n \psi$ with $\psi \in K[C]_P$ but there is no element $\psi' \in K[C]_P$ with $\varphi = (x-\alpha)^{n+1}\psi'$, then the *x-order* of φ, $\nu_{P,x}(\varphi)$, is defined to be n.

The x-order is called *good* if for $\varphi \in K[C]_P$, $\nu_{P,x}(\varphi) = 0$ implies that $\varphi(P) \neq 0$.

The *y-order* is defined analogously.

EXAMPLE Consider the function ring of the curve C: $x^3 + y^3 = 1$ defined over F_2 and the point P: $(10, 0)$ with coefficients in F_4.

The x-order of P is not good, because

$$(x-10)/y = (x-10)y^2/y^3 = (x-10)y^2/(x^3+1) = y^2/(x^2+10x+11) \quad (*)$$

and the denominator $x^2 + 10x + 11$ is non-zero at P. If the x-order were good, we would have to have $\nu_{P,x}((x-10)/y) > 0$ and therefore $\nu_{P,x}(1/y) \geq 0$. That would imply that $1/y$ lies in the local ring $K[C]_P$ of P, which cannot be, because $1/y$ has a pole at P.

It follows from Theorem 3.11 that the y-order at this point is good. Now let us check the discrete valuation axioms.

Since $1/y \notin K[C]_P$ we have $\nu_{P,x}(1/y) < 0$, and we have shown that $\nu_{P,x}(y) = 0$. Now $y(1/y) = 1 \in K[C]_P$ and we cannot represent 1 as $(x-10)^n\varphi$ with $n > 0$ and $\varphi \in K[C]_P$ because all such functions have value 0 at P. Hence $\nu_{P,x}(1) = 0 \neq \nu_{P,x}(y) + \nu_{P,x}(1/y)$, and axiom DV1 fails.

It is easy to check that all order functions satisfy axioms DV2 and DV3 (see the proposition below), so the only critical axiom is DV1. Verifying that $\nu_{P,y}$ satisfies axiom DV1 just involves following the steps of the following proposition.

PROPOSITION *Let C: $f(x,y) = 0$ be a curve and let P: $(x{=}\alpha, y{=}\beta)$ be a point of C with $\alpha \in K$. Then the x-order at P is good if and only if it is a discrete valuation on $K(C)$.*

Proof We shall write ν for $\nu_{P,x}$. We first check the claim made in the example that axioms DV2 and DV3 always hold. So let $\nu(\varphi) = m$ and $\nu(\psi) = n$. Then $\varphi = (x-\alpha)^m\varphi'$ and $\psi = (x-\alpha)^n\psi'$ with $\varphi, \psi \in K[C]_P$. Suppose, say, that $m \leq n$; then $\varphi' + (x-\alpha)^{n-m}\psi' \in K[C]_P$ and the equation $\varphi + \psi = (x-\alpha)^m(\varphi' + (x-\alpha)^{n-m}\psi')$ verifies that $\nu(\varphi+\psi) \geq m$ establishing DV2.

Certainly $\nu(x-\alpha) \geq 1$. It cannot be strictly greater than 1, because $(x-\alpha) = (x-\alpha)^2\varphi$ with $\varphi \in K[X]_P$ would imply that $\varphi = 1/(x-\alpha) \in K[C]_P$. That is impossible as $1/(x-\alpha)$ has a pole at P. So $\nu(x-\alpha) = 1$ and axiom DV3 holds.

Suppose that ν is good and $\nu(\varphi) = m$ and $\nu(\psi) = n$. Then $\varphi = (x-\alpha)^m\varphi'$ and $\psi = (x-\alpha)^n\psi'$ with $\varphi, \psi \in K[C]_P$. We can say further that $\nu(\varphi') = \nu(\psi') = 0$ for otherwise we could increase the power of $(x-\alpha)$ dividing φ or ψ. It follows that $\varphi'(P) \neq 0$ and $\psi'(P) \neq 0$. Hence $\varphi\psi = (x-\alpha)^{m+n}\varphi'\psi'$ and since $\varphi'\psi'(P) \neq 0$ we have $\nu(\varphi'\psi') = 0$. That implies that $(x-\alpha)^{m+n}$ is the highest power of $(x-\alpha)$ dividing $\varphi\psi$ with a quotient in $K[C]_P$ and thus $\nu(\varphi\psi) = m+n$ confirming axiom DV1.

The proof that axiom DV1 fails for bad orders follows the line of argument indicated in the example. Suppose that ν is bad. Then there exists a function

$\varphi \in K[C]_P$ such that $\nu(\varphi) = 0$ and $\varphi(P) = 0$. The second equation implies that $1/\varphi \notin K[C]_P$ and so $\nu(1/\varphi) < 0$. Then $1 = \varphi(1/\varphi)$, but $\nu(1) = 0 \neq 0 + \nu(1/\varphi)$ and so axiom DV1 fails. □

10.3 Valuation rings and places

Our main aim is to show that the places of F correspond to the possible discrete valuations of F over K and vice versa. Given a discrete valuation ν on F it is natural to consider the following ring.

DEFINITION Let ν be a discrete valuation on a field F; then the *valuation ring* V_ν is defined as the set of all $x \in F$ such that $\nu(x) \geq 0$ (including the zero element 0).

It is quite easy to show that V is a place ring.

PROPOSITION *Let ν be a K-valuation of the function field F over K. Then the valuation ring V_ν is a place.*

Proof By the definition of a K-valuation, every element $a \in K$ satisfies $\nu(a) = 0$ and so $a \in V$. Thus $K \subseteq V \subseteq F$. By DV3 there is a non-zero element $x \in F$ with $\nu(x) = 1$. This element x lies in V but not in K, so $K \subset V$. On the other hand it follows from DV1 that $\nu(1/x) = -1$ so $1/x \notin V$. Hence $V \subset F$ and we have established that V satisfies PL1.

Condition PL2 is almost trivial: if $x \notin V$, then by definition $\nu(x) < 0$. Hence by DV1 $\nu(1/x) > 0$ and so $1/x \in V$.

It remains to verify that V is closed under addition and multiplication. But if $x, y \in V$ then $\nu(x) \geq 0$ and $\nu(y) \geq 0$. By DV1 it follows that $\nu(xy) = \nu(x) + \nu(y) \geq 0$ and by DV2 it follows that $\nu(f + g) \geq \min\{\nu(x), \nu(y)\} \geq 0$. □

The converse statement that every place determines a unique discrete valuation is also true. For the special case of rational functions one can verify that directly, but the general proof is a little more involved.

EXAMPLE Rational functions over K.

For each irreducible polynomial $p(x) \in K[x]$ we define an order function. For a polynomial $f(x)$ the order $\nu_p(f)$ is the power n for which $f(x) = p(x)^n g(x)$ where $p(x)$ does not divide $g(x)$. For a rational function f/g we define $\nu_p(f/g) = \nu_p(f) - \nu_p(g)$.

PROPOSITION *Let $F = K(x)$ be the field of rational functions over the field K. Then the K-valuations on F are precisely the negative degree function and the order functions for the irreducible polynomials of $K[x]$.*

Proof It is straightforward to verify that the functions claimed are indeed K-valuations and that they are distinct. On the other hand, the proposition above shows every discrete valuation defines a place. In our case the places of $K(x)$ were determined in Example 9.5. They correspond to the irreducible polynomials in $K[x]$ and $1/x$. We must show that each place determines a unique valuation.

So let μ be a discrete valuation and let V be its valuation ring.

Suppose that V is the place ring corresponding to an irreducible polynomial $p(x) \in K[x]$ and denote the order function for $p(x)$ by ν. Then $1/p \notin V$ and so $\mu(p) > 0$. Further if $f(x)$ is a polynomial not divisible by p, then f and $1/f$ both lie in V and so by DV1, $\mu(f) = 0$. Thus for any polynomial of the form $g = p^n f$, where p does not divide f, we have $\mu(g) = n\mu(p) = \mu(p)\nu(g)$. It follows that $\mu = \mu(p)\nu$. Now by DV3 some function g must have $\mu(g) = \mu(p)\nu(g) = 1$ and it follows that $\mu(p) = 1$. Thus $\mu = \nu$.

Now suppose that V is the place ring at infinity. We let ν be the negative degree function. The function $z = 1/x$ lies in V, but its inverse x does not. Thus $\mu(z) > 0$. Hence all of the polynomials $K[z]$ lie in V. We can calculate $\mu(f)$ for a polynomial in $K[x]$ just as in Example 9.5. If $f = \sum a_i x^i$ has degree d, then $f(x) = \sum a_{d-i} z^i / z^d$. By assumption $a_d \neq 0$ and so $\mu(\sum a_{d-i} z^i) = \mu(a_d) = 0$. Therefore $\mu(f(x)) = -\deg(f)\mu(z)$. Hence for a rational function $f(x)/g(x)$ we have

$$\mu(f/g) = (\deg(g) - \deg(f))\mu(z) = \nu(f/g)\mu(z).$$

As before, the fact that there exists a function φ with $\mu(\varphi) = 1$ forces $\mu(z) = 1$. Thus $\mu = \nu$. □

10.4 Valuation rings determine valuations

The first step in showing that each place of a general function field determines a unique discrete valuation is to show that distinct valuations have different valuation rings.

PROPOSITION *Let μ and ν be distinct discrete valuations of a field F. Then the valuation rings V_μ and V_ν are distinct, indeed $V_\mu \not\subseteq V_\nu$.*

Proof Since $\mu \neq \nu$ there must exist $x \in F$ such that $\mu(x) \neq \nu(x)$. Replacing x by $1/x$ if necessary, we may assume that $\mu(x) \geq 0$. We choose x so that $\mu(x)$ is as small as possible.

If $\mu(x) = 0$ then both x and $1/x$ lie in V_μ. As $\nu(x) \neq \mu(x)$, one of x and $1/x$ does not lie in V_ν. So $V_\mu \not\subseteq V_\nu$ and the two rings are distinct. We therefore now assume that $\mu(x) = 0$ implies $\nu(x) = 0$.

If $\mu(x) > 1$, choose y such that $\mu(y) = 1$. By the choice of $\mu(y)$ it follows that $\nu(y) = \mu(y)$. But then $\mu(x/y) = \mu(x) - 1$ and $\nu(x/y) = \nu(x) - 1 \neq \mu(x) - 1$. That contradicts the choice of $\mu(x)$.

There remains the case that $\mu(x) = 1 \neq \nu(x)$. Choose y with $\nu(y) = 1$ and let $\mu(y) = n$. Then $\mu(y/x^n) = 0$ and so, by assumption, $\nu(y/x^n) = 0$. Thus

$$n\nu(x) = \nu(x^n) = \nu(y) = 1.$$

The only integer solutions of this equation are $n = \nu(x) = 1$ which is excluded by hypothesis and $n = \nu(x) = -1$. That implies x lies in V_μ but not in V_ν. Hence again $V_\mu \not\subseteq V_\nu$ and the two rings are distinct. □

10.5 Places are valuation rings

We are now ready to close the circle and show that every place determines a discrete valuation.

THEOREM *Let P be a place of a function field F over K; then there exists a K-valuation ν such that $F_P = V_\nu$.*

Proof Suppose that $F_P = V_\nu$ for some discrete valuation ν. Then certainly for $0 \neq x \in K$ the fact that x and $1/x$ both lie in F_P will force $\nu(x) = 0$ so ν will be a K-valuation. Furthermore if $x \in F_P$ has $\nu(x) = n$ we can construct a sequence $x = y_0, \ldots, y_n$ such that $y_i \in F_P$ and $\nu(y_i/y_{i+1}) = 1$ and so $y_i/y_{i+1} \in P$.

The idea of the proof is to define $\nu(x)$ as the greatest length of a sequence $x = y_0, \ldots, y_n$ with $y_i \in F_P$ and y_i/y_{i+1} in the ideal P and prove that it is a discrete valuation. We shall call these sequences *admissible*. The hardest step is to show that for fixed $x \neq 0$ the lengths of admissible sequences are bounded. If $x \in \bar{K}$, then $y \in F_P$ implies $y/x \in V$ and so $x/y \notin P$. For elements of $\bar{K}$ therefore n is always 0 and the claim is true.

Assume that $x \notin \bar{K}$. Then $|F : K(x)| = d$ is finite. We shall show that the length $n+1$ of any admissible sequence $x = y_0, \ldots, y_n$ is bounded by d. That follows once we have established that $y_0, \ldots, y_n$ are linearly independent over $K(x)$. So suppose that $\sum f_i y_i = 0$ with $f_i \in K(x)$ and not all f_i zero. Multiplying by a common denominator we can assume that f_i is a polynomial in x for all i and by eliminating common factors x we can assume that not all f_i are divisible by x. We denote by a_i the constant term $f_i(0)$ of f_i, which lies in K. It follows that $\sum a_i y_i = wx$ with $w \in V$. Let k be the largest index for which $a_k \neq 0$. Then $k \geq 1$ and

$$a_k = wx/y_k - \sum_1^{k-1} a_i y_i/y_k.$$

But all the terms x/y_k, y_i/y_k on the right hand side lie in P. Hence $a_k \in K \cap P = \{0\}$, a contradiction. Therefore $y_0, \ldots, y_n$ are linearly independent over $K(x)$ and so $n + 1 \leq |F : K(x)| = d$.

We can now define $\nu(x)$ for $0 \neq x \in F_P$ to be the largest n such that there is an admissible sequence $x = y_0, \ldots, y_n$. We first check that the axioms hold for elements of F_P, starting with DV3. If $x = y_0, \ldots, y_n$ is a longest admissible sequence starting with x, then $y_n \notin P$ while $y_i \in P$ for $i < n$. In particular $x \in P$ if and only if $\nu(x) > 0$. Further, if $x \in P$ a longest admissible sequence starting with y_{n-1} is y_{n-1}, y_n. So $\nu(y_{n-1}) = 1$.

Let x and y be two elements in F_P and let $y = y_0, \ldots, y_n$ be any admissible sequence for y. If $x/y \in F_P$ then $x = xy_0/y, \ldots, xy_n/y$ is an admissible sequence for x. So $\nu(x) \geq \nu(y)$. If $x/y \in P$ then $x, y = y_0, \ldots, y_n$ is an admissible sequence for x and therefore $\nu(x) > \nu(y)$. Thus $\nu(x) > \nu(y)$ if and only if $x/y \in F_P$ and $\nu(x) > \nu(y)$ if and only if $x/y \in P$.

Let $\nu(t) = 1$. Then it follows that $\nu(x) > 0$ if and only if $x/t \in F_P$. Thus $P = tF_P$. We can go further and say that if $\nu(x) = n$, then $x \in t^n F_P \setminus t^{n+1} F_P$.

Now we can verify axioms DV1 and DV2. Let $U = F_P \setminus P$. If $\nu(x) = m$ and $\nu(y) = n$, then $x/t^m \in U$ and $y/t^n \in U$ and so $xy/t^{m+n} \in U$. Thus $\nu(xy) = m + n$ establishing DV1.

Furthermore if, say, $m \leq n$, then both x and y lie in $t^m F_P$. Thus $x + y \in t^m F_P$ and so $\nu(x+y) \geq m$. Thus DV1 and DV2 hold for elements of F_P.

I leave it as an exercise (Exercise 1) for the reader to show that ν can be extended to the whole of F by setting $\nu(x/y) = \nu(x) - \nu(y)$ for $x, y \in F_P$ and that this extension satisfies the axioms for a discrete valuation. □

Elements t with $\nu_P(t) = 1$ are very useful calculation tools. So we give them a name and note a property that was established in the course of the proof.

DEFINITION An element $t \in P$ with $\nu_P(t) = 1$ is called a *local parameter* for the place P.

COROLLARY *If t is a local parameter for P and $\nu_P(x) = n$ then there exists $a \in F_P$ such that $\nu(x - at^n) > n$. The element a is uniquely determined modulo P.*

Proof The existence of a was established in the proof of the theorem. The uniqueness statement is left to Exercise 2. □

10.6 The approximation theorem

The major result of this chapter is Artin and Whaples' fundamental generalization of the Chinese remainder theorem (see Artin 1967). In the literature this theorem is also known as the theorem of independence.

THEOREM Approximation theorem (Artin–Whaples) *Let $P_1, \ldots, P_h$ be distinct places with valuation functions ν_i and suppose we are given $v_k \in F$ and an integer m_k for each place P_k. Then there exists $x \in F$ such that*

$$\nu_k(x - v_k) \geq m_k$$

for all $k = 1, \ldots, h$.

Remark The Chinese remainder theorem states that given a finite set of integer prime powers $p_1^{m_1}, \ldots, p_h^{m_h}$ (where the primes are distinct) and a finite set of target integers $v_1, \ldots, v_h$, there exists an integer x such that

$$x \equiv v_k \pmod{p_k^{m_k}}$$

for all $k = 1, \ldots, h$. It is proved by finding integers u_k such that

$$u_k \equiv 1 \pmod{p_k^{m_k}} \quad \text{and}$$
$$u_k \equiv 0 \pmod{p_j^{m_j}} \quad \text{for } j \neq k.$$

Then the required value x is $\sum u_k v_k$. The existence of these integers u_k is usually established by means of Euclid's algorithm.

The proof of the approximation theorem given here is based on the same idea with functions in place of integers, but the absence of Euclid's algorithm makes it harder to prove that the functions u_k exist.

Proof We shall first reduce the statement to the claim that certain elements of F exist, which are independent of the choices of v_k and m_k. Then we shall prove the existence of these elements by induction on the number of places h. The theorem is trivial for $h = 1$, so we shall assume throughout that $h \geq 2$. We choose n so that $n + \nu_k(v_j) \geq m_k$ for all j and k.

(1) It is sufficient to show that there exist $u_1, \ldots, u_h$ such that

$$\begin{aligned} \nu_k(u_k - 1) &\geq n \quad \text{and} \\ \nu_j(u_k) &\geq n \quad \text{for } j \neq k. \end{aligned}$$

For then with $x = \sum u_j v_j$ we have

$$\begin{aligned} \nu_k(x - v_k) &= \nu_k(\textstyle\sum u_j v_j - v_k) \\ &= \nu_k\big((u_k - 1)v_k + \sum_{j \neq k} u_j v_j\big) \\ &\geq \min\big(\{\nu_k(u_k - 1) + \nu_k(v_k)\} \cup \{\nu_k(u_j) + v_k(v_j) : j \neq k\}\big) \\ &\geq n + \min\{\nu_k(v_j)\} \\ &\geq m_k \end{aligned}$$

for all k.

(2) It is sufficient to establish the existence of $s_1, \ldots, s_h$ such that

$$\begin{aligned} \nu_k(s_k) &\geq n \quad \text{and} \\ \nu_j(s_k) &= 0 \quad \text{for } j \neq k. \end{aligned}$$

For, putting $t_k = \prod_{j \neq k} s_j$ we have

$$\begin{aligned} \nu_k(t_k) &= 0 \quad \text{and} \\ \nu_j(t_k) &\geq n \quad \text{for } j \neq k \text{ and hence} \\ \nu_j(t_k - s_k) &= 0 \quad \text{for all } j, k. \end{aligned}$$

Thus setting $u_k = t_k/(t_k - s_k)$ we have $u_k - 1 = s_k/(t_k - s_k)$ and so

$$\begin{aligned} \nu_k(u_k - 1) &= \nu_k(s_k) \geq n \quad \text{and} \\ \nu_j(u_k) &= \nu_j(t_k) \geq n \quad \text{for } j \neq k. \end{aligned}$$

(3) It is sufficient to find elements $r_1, \ldots, r_h$, such that

$$\begin{aligned} \nu_k(r_k) &> 0 \quad \text{and} \\ \nu_j(r_k) &< 0 \quad \text{for } j \neq k. \end{aligned}$$

For then

$$\nu_k(r_k - 1) = \nu_k(1) = 0 \quad \text{and}$$
$$\nu_j(r_k - 1) = \nu_j(r_k) < 0 \quad \text{for } j \neq k.$$

Hence setting $s_k = (r_k/(r_k - 1))^n$ we obtain

$$\nu_k(s_k) = \quad n\nu_k(r_k) \quad \geq n \quad \text{and}$$
$$\nu_j(s_k) = n(\nu_j(r_k) - \nu_j(r_k)) = 0 \quad \text{for } j \neq k.$$

(4) The existence of $r_1, \ldots, r_h$ when $h = 2$.
By Theorem 10.4, F_{P_1} is not a subset of F_{P_2} and F_{P_2} is not a subset of F_{P_1}. Choose $a \in F_{P_1} \setminus F_{P_2}$ and $b \in F_{P_2} \setminus F_{P_1}$. Put $r_1 = a/b$ and $r_2 = b/a$.

The existence of $r_1, \ldots, r_h$ when $h > 2$.
We assume that the existence has been established when there are $h - 1$ places and establish the existence of r_1. The existence of r_k for $k = 2, \ldots, h$ follows by symmetry.

Applying our induction hypothesis to $1, 3, \ldots, h$ we may assume that we have a such that

$$\nu_1(a) > 0 \quad \text{and}$$
$$\nu_j(a) < 0 \quad \text{for } j > 2.$$

If $\nu_2(a) < 0$, we may take $r_1 = a$, so we assume that $\nu_2(a) \geq 0$. Now choose b such that $\nu_1(b) > 0$ and $\nu_2(b) < 0$ and choose $l \geq 0$ such that $l\nu_k(a) + \nu_k(b) < 0$ for all $k > 2$.

If $\nu_2(a) = 0$, set $r_1 = a^l b$. Then we have

$$\nu_1(r_1) = l\nu_1(a) + \nu_1(b) > 0$$
$$\nu_2(r_1) = \nu_2(b) \qquad < 0 \quad \text{and}$$
$$\nu_j(r_1) = l\nu_j(a) + \nu_j(b) < 0 \quad \text{for } j > 2.$$

If on the other hand $\nu_2(a) > 0$, then

$$\nu_j(1 + a^{-l}) = -\nu_j(a^l) \quad \text{for } j = 1, 2 \text{ and}$$
$$\nu_j(1 + a^{-l}) = 0 \qquad \text{otherwise.}$$

Hence putting $r_1 = a^l b(1 + a^{-l})$ we have

$$\nu_1(r_1) = \nu_1(b) \qquad > 0$$
$$\nu_2(r_1) = \nu_2(b) \qquad < 0 \quad \text{and}$$
$$\nu_j(r_1) = l\nu_j(a) + \nu_j(b) < 0 \quad \text{for } j > 2.$$

□

COROLLARY *With the notation of the theorem, given any sequence of integers* $n_1, \ldots, n_h$ *there exists an element* $x \in F$ *with* $\nu_i(x) = n_i$ *for all* $i = 1, \ldots, h$.

Proof For each i let t_i be a local parameter for P_i, so that $\nu_i(t_i) = 1$. By the theorem there exists an element $x \in F$ such that

$$\nu_k(x - t_i^{n_i}) > n_i \quad \text{for } i = 1, \ldots, h.$$

It follows that $\nu_i(x) = n_i$ for all $i = 1, \ldots, h$. □

Remark Neither the theorem nor its corollary holds if the set of places is infinite. That is a consequence of the degree theorem proved in the next chapter.

10.7 Places of F over a given place of $K(x)$

It follows from the results of Chapter 9 that for a non-constant $x \in F$ every place of F is an *extension* of a place of $K(x)$. As a first consequence of the approximation theorem, we shall show that a single place of $K(x)$ has only finitely many extensions.

PROPOSITION *Assume that* $x \in F$ *is not a constant, and that* P *is a place of* F. *Let* U *be the set of place units* $F_P \setminus P$. *Then either* $K(x) \subset U \cup \{0\}$ *or* $P \cap K(x)$ *is a place of* $K(x)$. *Conversely, to every place* Q *of* $K(x)$ *there exists a place* P *of* L *with* $P \cap K(x) = Q$. *The number of such places is at most equal to* $|F : K(x)|$.

Proof The existence of a place P extending a place Q of $K(x)$ follows directly from Theorem 9.7. Conversely, let P be a place of F. Then $W = F_P \cap K(x)$ satisfies PL2. If $K(x) = W$, then all the non-zero elements of $K(x)$ are invertible in F_P. On the other hand, if $W \neq K(x)$ it also satisfies PL1. So it is a place of $K(x)$.

To prove that a place Q of $K(x)$ has only finitely many extensions, we assume that Q corresponds to an irreducible polynomial $p(x)$. Otherwise Q is the place at infinity. In that case we can replace x by $1/x$ and Q becomes the place corresponding to x. The assumption implies that the place ring $K(x)_Q$ of Q contains $K[x]$ and so the same holds for all place rings extending Q.

Let $P_1, \ldots, P_n$ with valuations ν_i be a finite set of places such that $P_i \cap K(x) = Q$ and let t_i be elements such that $\nu_i(t_j) = 1$ if $i = j$ and $\nu_i(t_j) = 2$ otherwise. Such elements exist by the corollary to the approximation theorem. We shall show that the elements t_i are linearly independent over $K(x)$.

Suppose therefore that $\sum f_i(x)t_i = 0$ with $f_i(x) \in K(x)$ not all zero. Multiplying by a common denominator, we may assume that $f_i(x) \in K[x]$ for all i. For the irreducible polynomial $p(x)$ corresponding to Q let $f_i(x) = p(x)g_i(x) + h_i(x)$ in $K[x]$ where $\deg h_i(x) < \deg p(x)$. Dividing by a common factor $p(x)^r$, we may assume that not all $h_i(x)$ are zero, say $h_1(x) \neq 0$. Then

$$p(x) \sum g_i(x)t_i = -\sum h_i(x)t_i.$$

Since $K[x]$ is contained in the place ring F_{P_1}, it follows that $\nu_1(g_i(x)) \geq 0$ and hence that
$$\nu_1\left(p(x)\sum g_i(x)t_i\right) > 1.$$
On the other hand, $\deg h_i(x) < \deg p(x)$ implies that $h_i(x) \notin P_1$. It follows that $\nu_1(h_1(x)t_1) = 1$ while $\nu_1(h_i(x)t_i) \geq 2$ for $i = 2, \ldots, n$. Hence
$$\nu_1\left(\sum h_i(x)t_i\right) = 1.$$

This contradiction shows that the set $\{t_1, \ldots, t_n\}$ is linearly independent over $K(x)$.

It follows that there cannot be more places extending Q than $|F : K(x)|$, proving the claim. □

10.8 Exercises

1. Conclude the proof of Theorem 10.5 by showing that the function ν defined on the place ring F_P can be uniquely extended to a function on $F\backslash\{0\}$, by setting $\nu(x/y) = \nu(x) - \nu(y)$ for $x, y \in F_P$ and that this extension satisfies the axioms for a discrete valuation.
2. Let t be a local parameter for the place P and let $\nu_P(x) = n$. Show that if $\nu_P(x - at^n) > n$ for some $a \in F$, then $\nu_P(x - bt^n) > n$ for $b \in F$ if and only if $a - b \in P$.
3. Find an integer x such that
$$\begin{aligned} x &\equiv 7 \pmod{9} \\ x &\equiv 11 \pmod{32}. \end{aligned}$$
4. Find a local parameter of the local ring of the point $(0, 1)$ in the function ring of the curve C: $x^3 + y^3 = 1$ defined over F_2. Do the same for the point $(0, 0)$ of the Klein quartic.
5. For the singular point $(0, 0)$ in Exercise 9.3 find local parameters for each place extending the local ring of the point. Do the same for the singular point $(0, 0)$ in Exercise 9.4.
6. Find a function in the function field of the Klein quartic over F_2 that has a pole of order 1 at $(0, 0)$ (so its inverse is a local parameter), a zero of order 1 at $(3, 10)$ (so it is a local parameter), and has a finite non-zero value at $(6, 8)$.
7. Show that it is not possible to find a function in the function field of the Klein quartic that has a zero of order 1 at $(3, 10)$ and a zero of order 2 at $(5, 11)$. Why does that not contradict the approximation theorem? Where would the argument of the approximation theorem fail if it were used to try and prove the existence of such a function?

11

DIVISORS

In this chapter we shall prove the degree theorem and Riemann's theorem. For that we need to generalize the concept of the degree of a point to places and then extend the concept of a divisor to places in the natural way.

11.1 The degree of a place

EXAMPLE Consider the function ring of the curve C: $x^3 + y^3 = 1$ defined over F_2 and the point P: $(10, 0)$ with coefficients in F_4.

In Example 10.2 we calculated

$$(x - 10)/y = (x - 10)y^2/y^3 = (x - 10)y^2/(x^3 + 1) = y^2/(x^2 + 10x + 11).$$

This calculation is performed over F_4, but what happens over F_2? There we can perform an analogous calculation using the minimal polynomial $x^2 + x + 1$ of 10 over F_2 in place of $x - 10$:

$$(x^2 + x + 1)/y = (x^2 + x + 1)y^2/(x^3 + 1) = y^2/(x + 1).$$

This lies in the local ring $V = \mathsf{F}_2[C]_P$ of P over F_2. Thus V consists of those rational functions with non-negative y-order and y is a local parameter. The maximal ideal of V is yV. Every element of V/yV can be represented by a quotient $g(x)/h(x)$ of polynomials in x alone, where $h(P) \neq 0$. That is equivalent to the condition that the minimal polynomial $x^2 + x + 1$ of 10 over F_2 does not divide $h(x)$. That polynomial is equal to $y^3/(x + 1)$ in V and so it represents 0 in V/yV. On the other hand a polynomial in x that is not divisible by $x^2 + x + 1$ is non-zero at P and so does not represent 0. It follows that $V/yV \cong \mathsf{F}_2[x]/(x^2 + x + 1) = \mathsf{F}_4$. In this way we can find the degree of the point P using only its local ring over F_2.

It is not difficult to extend the argument of this example and show the degree of the simple point P can always be read off from its local ring V with maximal ideal tV. The degree is $|V/tV : K|$ (see Exercise 1). That suggests that we should define the degree of a place P of an arbitrary function field F as $|F_P/P : K|$. In order for the definition to make sense, it is first necessary to prove that $|F_P/P : K|$ is finite.

PROPOSITION *Let V be a place ring of F with maximal ideal $P = tV$. Then $|F : K(t)|$ is finite, and $|\bar{K} : K| \leq |V/P : K| \leq |F : K(t)|$.*

Proof We have established in Corollary 9.7b that $\bar{K} \subseteq V \setminus P$. Hence $\bar{K} \subseteq V/P$, establishing the first inequality. Since $t^{-1} \notin V$, it follows that $t \notin \bar{K}$. Hence by Proposition 9.2, $|F : K(t)|$ is finite.

As K is a subset of $\bar{K}$ it follows that it is a subset of V/P so V/P is a vector space over K. Let $u_1, \ldots, u_d \in V$ be such that the residues $\bar{u}_1, \ldots, \bar{u}_d$ modulo P are linearly independent over K. Suppose that

$$\sum_{i=1}^{d} f_i(t) u_i = 0,$$

where $f_i(t) \in K(t)$. Multiplying by a common denominator, we may assume that $f_i(t)$ is a polynomial in $K[t]$. Letting a_i be the constant term of $f_i(t)$, we note that the image of $f_i(t) \in V/P$ is a_i. Hence

$$\sum_{i=1}^{d} a_i \bar{u}_i = 0.$$

It follows that $a_i = 0$ for all i. Thus t divides $f_i(t)$ for all t. Repeating the argument, we find that t^n divides the polynomial $f_i(t)$ for all positive powers n. That is only possible if $f_i(t) = 0$ for all i. So $u_1, \ldots, u_d$ are linearly independent over $K(t)$ and $|F : K(t)| \geq d$. Hence $|V/P : K| \leq |F : K(t)|$. □

Having shown that $|V : K|$ is always finite, we can define it as the degree of a place.

DEFINITION The *degree* of the place P, denoted by $d(P)$, is the degree of the field extension $|F_P/P : K|$. A place is called *rational* if $d(P) = 1$, that is $F_P/P = K$.

With this definition we have the following immediate corollary of the proposition. It establishes a simple criterion for the base field K of a function field to be the full field of constants $\bar{K}$. The condition is not necessary, but it is satisfied in all the examples that will be of interest to us.

COROLLARY *If the function field F defined over K has a rational place, then $\bar{K} = K$ and hence the constant degree $\kappa = |\bar{K} : K|$ is 1.* □

11.2 Divisors

We generalize the concept of a divisor from Chapter 4 by replacing conjugacy classes of points by places. Just as in that chapter, divisors are a convenient tool for specifying conditions for several places at a time.

DEFINITION A *divisor* D of a function field F assigns an integer value $D(P)$ to every place of F, such that $D(P) = 0$ for all but finitely many places. We denote the divisor by a formal sum $\sum D(P)P$. The set of places for which $D(P) \neq 0$ is called the *support* of D.

The following operations are defined on divisors $A = \sum A(P)P$ and $B = \sum B(P)P$:

$$A + B = \sum (A(P) + B(P))P,$$
$$A - B = \sum (A(P) - B(P))P,$$
$$A \cup B = \sum \max\{A(P), B(P)\}P, \text{ and}$$
$$A \cap B = \sum \min\{A(P), B(P)\}P.$$

The first two operations make the set of divisors into an *abelian group*. The second two make the set into a *lattice*. Although the operations will be useful, we shall make no use of either group theory or lattice theory.

The *null divisor*, which will always be denoted by $N = \sum 0P$, is the zero of the additive group. We shall say that $A \leq B$ if $A(P) \leq B(P)$ for all P. If $N \leq D$, then D is called *positive* or *effective*.

The theorems from Chapter 4 that we aim to prove concern L-spaces and so we naturally also extend their definition in the obvious way.

DEFINITION Let S be a non-empty set of places.

We define $L(D, S)$ to be the set of $x \in F$ such $\nu_p(x) + D(P) \geq 0$ for all $P \in S$. These spaces are called *L-spaces*. If S is the set of all places the vector space $L(D, S)$ is denoted by $L(D)$. It is called the *space of functions over D*.

The dimension of $L(D, S)$ is denoted by $l(D, S)$ and the dimension of $\dim L(D)$ is denoted by $l(D)$ and called the *rank* of the divisor D.

The following proposition shows that the approximation theorem yields a lot of information about L-spaces.

PROPOSITION *The set $L(D, S)$ is a vector space over K. If S is finite then $l(D, S)$ is infinite.*

Proof The verification of the vector space properties is straightforward and left as an exercise (Exercise 2).

Suppose that $S \neq \emptyset$ is finite. Given an integer n, it follows from the corollary to the approximation theorem that there exists an element $x_n \in F$ such that $\nu_P(x_n) = n$ for all $P \in S$. For any finite set T of integers and any linear combination $y = \sum_{n \in T} a_n x_n$ with $0 \neq a_i \in K$, we have $\nu_P(y) = \min\{n : n \in T\}$ for all $P \in S$. Thus y cannot be zero and so the elements x_n are linearly independent over K. For $n \geq \max\{-D(P) : P \in S\}$, the element x_n lies in $L(D, S)$. Hence $l(D, S)$ is infinite. □

11.3 Relative dimensions

Although the spaces $L(D, S)$ may have infinite dimension. The relative dimension of $L(B, S)/L(A, S)$ for divisors $A \leq B$ is always finite. Recall that the relative dimension of L_1/L_2 for two vector spaces over a field K is the dimension of the factor space. It is the size d of a *(relative) basis of* L_1 *modulo* L_2. That is a set of elements $u_1, \dots, u_d \in L_1$ such that

(a) Every element $v \in L_1$ can be written as a linear combination $\sum a_i u_i + z$ with $a_i \in K$ and $z \in L_2$. We say $u_1, \dots, u_n$ *span* L_1 modulo L_2.
(b) The only linear combination $\sum a_i u_i$ that lies in L_2 has $a_i = 0$ for all $i = 1, \dots, d$. We say $u_1, \dots, u_n$ are *linearly independent* modulo L_2.

If $L_2 = \{0\}$, then these definitions revert to the standard definitions of spanning and linear independence. So a basis modulo $\{0\}$ is just an ordinary basis.

Not only is the relative dimension of $L(B,S)/L(A,S)$ finite, but it satisfies a simple formula.

PROPOSITION *Let $A \leq B$ be divisors and S be any finite set of places. Then $L(A,S) \subseteq L(B,S)$ and*

$$\dim L(B,S)/L(A,S) = \sum_{P \in S} (B(P) - A(P))d(P).$$

Proof The fact that $L(A,S) \subseteq L(B,S)$ follows directly from the definition.

To establish the formula we prove the case where $B = A + Q$ for a single place $Q \in S$. The inductive argument that then proves the general case is left as an exercise (Exercise 4).

To determine the dimension we let $u \in F$ satisfy $\nu_P(u) = -B(P)$ for all $P \in S$ (such a u exists by the corollary to the approximation theorem). Then $\nu_Q(u) + A(Q) < 0$, so $u \notin L(A,S)$.

Now let $v_1, \dots, v_d \in F_Q$ be chosen such that the residues $\bar{v}_1, \dots, \bar{v}_d$ form a basis of F_Q/Q over K (thus $d = d(Q)$). By the approximation theorem there exist $x_1, \dots, x_d$ such that

$$\nu_Q(x_i - v_i) \geq 1 \text{ and}$$
$$\nu_P(x_i) \geq 0 \text{ for } P \in S \setminus \{Q\}.$$

Since $\nu_Q(v_i) = 0$, it follows that $\nu_P(x_i) \geq 0$ for all $P \in S$ and hence $x_i u \in L(B,S)$. We shall show that the elements $x_i u$ form a basis of $L(B,S)$ modulo $L(A,S)$.

To show that every element of $L(B,S)$ can be represented as a linear combination $\sum a_i x_i u + zu$ where $zu \in L(A,S)$, let $w \in L(B,S)$. Then $\nu_Q(w/u) \geq 0$ and hence $w/u \in F_Q$. Therefore $w/u = \sum a_i x_i + z$ with $z \in Q$. As $z \in Q$, we have $\nu_Q(z) > 0$. For $Q \neq P \in S$, we have $\nu_P(w/u) \geq 0$ and $\nu_P(\sum a_i x_i) \geq 0$. From the equation $z = w/u - \sum a_i x_i$ it follows that $\nu_P(z) \geq 0$. Hence $zu \in L(A,S)$. The fact that $w = \sum a_i x_i u + zu$ now establishes the claim.

Finally we must show that the elements $x_1 u, \dots, x_d u$ are linearly independent modulo $L(A,S)$. So suppose now that $\sum a_i x_i u \in L(A,S)$; then $\nu_Q(\sum a_i x_i) + \nu_Q(u) \geq -A(Q)$. By its construction $-\nu_Q(u) = B(Q) = A(Q) + 1$.

Hence $\nu_Q(\sum a_i x_i) \geq 1$, or in other words $\sum a_i x_i \in Q$. That implies $\sum a_i \bar{x}_i = \bar{0}$ in F_Q/Q, but $x_1, \ldots, x_d$ are assumed to be a basis of F_Q/Q. Hence the coefficients satisfy $a_1 = \cdots = a_d = 0$. This proves the claim and establishes that $x_1 u, \ldots, x_d u$ form a basis of $L(A+Q,S)$ modulo $L(A,S)$. Hence $\dim L(A+Q,S)/L(A,S) = d = d(Q)$ as required. □

The proposition suggests that we make the following definition.

DEFINITION The *degree* of the divisor D is $d(D) = \sum D(P)d(P)$, where $d(P)$ denotes the degree of the place P.

The proposition then has the following useful corollary.

COROLLARY *Let $A \leq B$ be divisors and let S be a finite set of places. Then $L(A,S) \subseteq L(B,S)$ and $\dim L(B,S)/L(A,S) \leq d(B) - d(A)$, with equality if S contains supports of both A and B (that is, all places P where $A(P) \neq 0$ or $B(P) \neq 0$).* □

11.4 Elements of F as functions on places

It is now possible to justify the name 'field of algebraic functions' by interpreting the elements of F as functions on places.

DEFINITION Let $x \in F$ and P be a place with valuation ν_P. If $x \in F_P$ we say the *value* $x(P)$ of x at P is the image of x in F_P/P. Otherwise we say the value of x at P is ∞. We say P is a *zero* of x if $x \in P$ (that is equivalent to saying that the value of x is 0 or $\nu_P(x) > 0$). We say P is a *pole* of x if $x \notin F_P$ (that is equivalent to saying that the value of x is ∞ or $\nu_P(x) < 0$).

Notice that the value of x at P may lie in an extension field of K. If the place is rational, the value lies in $F_P/P = K$. This accords nicely with our examples.

We can now prove the zeros theorem which states that the number of zeros of $x \in F$ is bounded by the degree $|F : K(x)|$. Later, in the proof of the degree theorem we shall show that this bound is always exact if the zeros are weighted by the degree of the place.

THEOREM Zeros theorem *Let $0 \neq x \in F$ and let S be the set of zeros of x. Then S is empty if and only if x is a constant and*

$$\sum_{P \in S} d(P)\nu_P(x) \leq |F : K(x)|. \qquad (*)$$

Proof For a constant the statement is trivial. So assume that x is not constant. By Corollary 9.7a there exists a place P containing x. Then $P \in S$ and so S is not empty. Let T be any finite subset of S. We shall show that the bound $(*)$ holds for T. Then it follows that such a set cannot have more than $|F : K(x)|$ elements. So S is itself finite and the bound holds for S.

Let $D = \sum_{P \in T} \nu_P(x) P$ (so $D(Q) = 0$ for any place $Q \notin T$). As T contains only zeros of x, it is immediate that D is positive. So $D \geq N$, where N is the null divisor $N(P) = 0$ for all P. Then by Proposition 11.3, $L(D,T)/L(N,T)$

has dimension $r = \sum_{P \in T} D(P)d(P)$. Let $z_1, \ldots, z_r$ form a basis of $L(D,T)$ modulo $L(N,T)$. We shall show that $z_1, \ldots, z_r$ are linearly independent over $K(x)$ establishing the inequality.

For suppose that $\sum f_i(x)z_i = 0$ for some non-zero elements $f_i(x) \in K(x)$. As usual we may multiply by a common denominator and divide by any common factor x^n. So we can assume that $f_i(x) \in K[x]$ for all i and that not all $f_i(x)$ are divisible by x.

Writing $f_i(x) = a_i + xg_i(x)$ we get

$$\sum a_i z_i = -x \sum g_i(x) z_i. \qquad (**)$$

By assumption, $\nu_P(x) > 0$ for $P \in T$ and hence $\nu_P(g(x)) \geq 0$. Therefore $\nu_P \sum g_i(x)z_i \geq -D(P) = -\nu_P(x)$ and thus $\nu_P(x \sum g_i(x)z_i) \geq 0$. That implies that the right hand side of eqn (**) lies in $L(N,T)$. On the other hand, $z_1, \ldots, z_r$ are assumed to be linearly independent modulo $L(N,T)$. It follows that $a_i = 0$ for all i. But that implies that x divides $f_i(x)$ for all i contradicting our assumptions. Hence the elements $z_1, \ldots, z_r$ are linearly independent over $K(x)$ and so $r \leq |F : K(x)|$ as claimed. □

This theorem will have important consequences in the next few paragraphs, and indeed is the key to the proof of the degree theorem. An immediate consequence is that distinct elements of F define different functions on places.

COROLLARIES (a) *If $x \neq 0$ then x has only finitely many zeros.*
(b) *Let x and y be elements of F; then if $x(P) = y(P)$ for all places P, it follows that $x = y$.*

Proofs (a) If x is a constant then it has no zeros. Otherwise $|F : K(x)| = d$ is finite, and so by the theorem, x cannot have more than d zeros.

(b) The condition implies that $\nu_P(x - y) > 0$ for all places P, so all places are zeros of $x - y$. By Corollary 9.7c there are infinitely many places and so $x - y$ has infinitely many zeros. Then it follows from Corollary (a) that $x - y = 0$. □

11.5 The space $L(D)$

In the next two paragraphs we shall prove that every divisor has finite rank $l(D)$. We begin with two easy special cases.

PROPOSITION (a) *For the null divisor N we have $L(N) = \bar{K}$, and so $l(N) = \kappa$.*
(b) *For any divisor $D < N$, we have $L(D) = \{0\}$.*

Proof (a) If $0 \neq x \in L(D)$ for $D \leq N$, then $\nu_P(x) \geq 0$ for all places P. It follows that $1/x$ has no zeros. Therefore $1/x$ is a constant. Since any constant lies in $L(N)$, that establishes the first claim.

(b) We have already shown that every element of $L(D)$ is a constant. If $D < N$, then $D(P) < 0$ for some P. The only constant x with $\nu_P(x) + D(P) \geq 0$ is 0. Hence $L(D) = \{0\}$. □

11.6 Finiteness of the rank of a divisor

We shall show that the function $l(D) - d(D)$ decreases as D increases. Then from the fact that the value is determined for divisors $\leq N$ we can find an upper bound for the function.

PROPOSITION *Let $A \leq B$ be divisors. Then $l(A) - d(A) \geq l(B) - d(B)$.*

Proof Before starting the proof recall a standard result on vector spaces (considered as abelian groups).

THEOREM Isomorphism theorem *Let $U, V \subseteq W$ be vector spaces over a field. Then*

$$U/U \cap V \cong (U + V)/V. \qquad \square$$

If $l(A)$ is infinite there is nothing to prove, so we may assume that $l(A) < \infty$. Let S be a finite set of places including the supports of A and B, so that $A(P) = B(P) = 0$ for any $P \notin S$. Then $L(A) = L(B) \cap L(A, S)$ and so by the isomorphism theorem

$$\begin{aligned} L(B)/L(A) &= L(B)/\big(L(B) \cap L(A, S)\big) \\ &\cong \big(L(B) + L(A, S)\big)/L(A, S) \\ &\subseteq L(B, S)/L(A, S). \end{aligned}$$

By Proposition 11.3, the last vector space has finite dimension $d(B) - d(A)$. Hence $L(B)/L(A)$ is finite dimensional. It follows that $L(B)$ is finite dimensional. In that case the dimension of $L(B)/L(A)$ is $l(B) - l(A)$. Combining the two statements we obtain $l(B) - l(A) \leq d(B) - d(A)$, establishing the inequality. $\square$

COROLLARIES (a) *$l(D)$ is finite for all divisors D.*
(b) *If D is positive, then $l(D) \leq d(D) + \kappa$, where $\kappa = |\bar{K} : K|$.*

Proofs (a) Let $A = D \cap N$, so that $A(P) = \min\{D(P), 0\}$ for all places P. Then $A \leq N$ so $L(A) \subseteq \bar{K}$ and hence $l(A) \leq \kappa$. By the proposition, $l(D) \leq d(D) - d(A) + l(A)$, which is finite.

(b) Comparing the given divisor with the null divisor N it follows from the proposition that

$$l(D) - l(N) \leq d(D) - d(N).$$

The claim then follows from the established values $d(N) = 0$ and $l(N) = \kappa$. $\square$

11.7 The divisor of zeros of a function

In view of Corollary 11.4a we make the following definition.

DEFINITION For a non-zero element $x \in F$ the *divisor of zeros* $[x]_0$ of x is defined by $[x]_0(P) = \max\{\nu_P(x), 0\}$ for all places P.

We also define the *divisor of poles* $[x]_\infty$ to be the divisor of zeros of x^{-1}. So the support of $[x]_0$ consists of the zeros of x, while the support of $[x]_\infty$ consists of the zeros of x^{-1}, which are the poles of x. Both divisors are positive.

The zeros theorem can now be restated in the concise form

$$d[x]_0 \leq |F : K(x)|.$$

It is useful to be able to calculate the divisors of poles and zeros of polynomials in a non-constant x. The divisor of poles of a polynomial is easily determined if the divisor of poles of x is known, but the divisor of zeros of $f(x)$ is not closely related to the divisor of zeros of x.

PROPOSITION (a) *Let x be a non-constant element of F and let $f(x)$ be a polynomial in $K[x]$. Then the divisor of poles of $f(x)$ is $[f(x)]_\infty = \deg(f)[x]_\infty$.*
(b) *If a place P is not a pole of x, then there exists a polynomial $t(x)$ with a zero at P.*
(c) *If $f(x)$ and $g(x)$ are polynomials, then the highest common factor (f,g) of f and g has divisor of zeros $[(f,g)]_0 = [f]_0 \cap [g]_0$.*

Proof (a) Write $f(x) = \sum a_i x^i$ and consider the terms of the sum. For any place P we have $\nu_P(a_i x^i) = i\nu_P(x)$ (unless $a_i = 0$), and $\nu_P(f) \geq \min\{\nu_P(a_i x^i)\}$. Thus if $\nu_P(x) \geq 0$, it follows that $\nu_P(f) \geq 0$. If, on the other hand, $\nu_P(x) < 0$, then all the non-zero terms of the sum have distinct values and so $\nu_P(f) = \min\{i\nu_P(x) : a_i \neq 0\}$. This value is $\deg(f)\nu_P(x)$, establishing the first statement.

(b) The intersection $P \cap K(x)$ is a place Q of $K(x)$ and Q is not the place at infinity, because P is not a pole of x. Hence Q corresponds to an irreducible polynomial $t(x)$. Then $t(P)$ is zero proving the second statement.

(c) Let $h = (f,g)$. Then $f(x) = q(x)h(x)$ for some polynomial $q(x)$. So $\nu_P(f) = \nu_P(q) + \nu_P(h)$ for all places P. From the first part of the proof it follows that $\nu_P(q)$ can only be negative at a pole of x. There $\nu_P(h)$ cannot be positive and so $\nu_P(f) \geq \nu_P(h)$ for all zeros of h. That implies that $[h]_0 \leq [f]_0$, and by symmetry $[h]_0 \leq [g]_0$. So $[h]_0 \leq [f]_0 \cap [g]_0$.

As x is a non-constant, $K[x]$ is a true polynomial ring. So we can write $h(x) = s(x)f(x) + t(x)g(x)$ for certain polynomials $s(x)$ and $t(x)$. If P is not a pole of x, $\nu_P(s) \geq 0$ and $\nu_P(t) \geq 0$. Thus $\nu_P(sf) \geq \nu_P(f)$ and $\nu_P(tg) \geq \nu_P(g)$. Hence

$$\nu_P(h) \geq \min\{\nu_P(sf), \nu_P(tg)\} \geq \min\{\nu_P(f), \nu_P(g)\}.$$

So $[h]_0 \geq [f]_0 \cap [g]_0$. □

The proposition allows us to find special bases of F over $K(x)$, which will be used in the proof of Riemann's theorem.

COROLLARIES *Let x be a non-constant.*
(a) *For any $y \in F$ there exists a non-zero element of $g(x) \in K[x]$ such that the poles of $g(x)y$ lie among the poles of x.*
(b) *F has a basis $w_1, \ldots, w_d$ over $K(x)$ such that the poles of w_i lie among the poles of x for all w_i.*
(c) *F has a basis $w_1, \ldots, w_d$ over $K(x)$ such that the poles of w_i lie among the zeros of x for all w_i.*

Proofs (a) Let S be the set of poles of y that are not poles of x. For each $P \in S$ let $t_P(x)$ be a polynomial in x with a zero at P. Let $g(x) = \prod_{P \in S} t_P(x)^{-\nu_P(y)}$. Then by construction, $g(x)y$ has no poles in S. The poles of this product lie among the poles of $g(x)$ and the other poles of y. By the proposition the poles of $g(x)$ lie among the poles of x, and by construction the poles of y that do not lie in S also lie among the poles of x.

(b) Apply Corollary (a) to each of the elements of an arbitrary basis $y_1, \ldots, y_n$, obtaining $w_1 = g_1(x)y_1, \ldots, w_n = g_n(x)y_n$ such that the poles of w_i lie among the poles of x. As the elements of $K(x)$ act as scalars multiplying basis elements by non-zero terms $g_i(x)$ does not affect the basis property.

(c) Apply Corollary (b) to x^{-1} and observe that $K(x^{-1}) = K(x)$. □

11.8 A lower bound for the rank of a divisor of zeros

The next proposition is a special case of Riemann's theorem.

PROPOSITION (Riemann) *Let $x \in F$ be a non-constant and $n = |F : K(x)|$. Then there exists an integer constant $t \in \mathsf{Z}$ such that for all $m \geq 0$ we have $l(m[x]_0) \geq mn - t$.*
Proof We need only prove the existence of t for sufficiently large values m, because t can always be adjusted to account for a finite number of special cases. So let $w_1, \ldots, w_n$ be a basis of F over $K(x)$ such that the poles of w_i lie among the zeros of x (such a basis exists by Corollary 11.7c). Let S be the set of zeros of x and choose s so that $\nu_P(w_i) \geq -s\nu_P(x)$ for all $i = 1, \ldots, n$ and all $P \in S$. We shall show that for $m \geq s$ the inequality holds with $t = sn$.

To do that we must find a set of $mn - sn$ linearly independent elements of $L(m[x]_0)$. The elements we choose are the fractions w_i/x^r for $r = 0, \ldots, m - s$.

For each $P \in S$

$$\nu_P(w_i/x^r) = \nu_P(w_i) - r\nu_P(x) \geq -(s+r)\nu_P(x) \geq -m\nu_P(x).$$

For $P \notin S$ we have $\nu_P(x) \leq 0$ and $\nu_P(w_i) \geq 0$ and hence $\nu_P(w_i/x^r) \geq 0$. Thus $w_i/x^r \in L(m[x]_0)$.

By assumption $w_1, \ldots, w_n$ are linearly independent over $K(x)$. Furthermore, $1/x^0, \ldots, 1/x^{m-s}$ lie in $K(x)$ and they are linearly independent over K, because $1/x$ is not a constant. Hence the elements w_i/x^r for $i = 1, \ldots, n$ and $0 \leq r \leq m - s$ are linearly independent over K. As they lie in $L(m[x]_0)$ it follows that $l(m[x]_0) \geq mn - sn$. □

11.9 The degree theorem

It is now possible to prove the degree theorem, which is a consequence of the following expansion of the zeros theorem.

THEOREM *Let x be a non-constant function; then x has only finitely many zeros and poles and if $[x]_0$ is the divisor of zeros of x, then $d([x]_0) = |F : K(x)|$.*

Proof Let $Z = [x]_0$ and $n = |F : K(x)|$. It follows from the zeros theorem (Theorem 11.4) that x has only finitely many zeros and that $d(Z) \leq n$. We shall estimate $l(mZ)$ in two ways to establish equality.

By Corollary 11.6b we have $l(mZ) \leq md(Z) + \kappa$ for all positive m. On the other hand Proposition 11.8 states that $mn - t \leq l(mZ)$ for some integer t and all positive m. So

$$mn - t \leq md(Z) + \kappa.$$

Dividing by m we obtain

$$n - t/m \leq d(Z) + \kappa/m.$$

It follows that for $m > t + \kappa$ we have $n - 1 < d(Z) \leq n$. Thus $d(Z) = n$. □

COROLLARY Degree theorem *The sum* $\sum \nu_P(x)d(P)$ *has only finitely many terms and its value is zero.*

Proof Let S be the set of zeros of x and T be the set of poles of x. Let Z be the divisor of zeros of x and let Y be the divisor of zeros of $1/x$. Then for any $P \notin S \cup T$ we have $\nu_P(x) = 0$. Thus

$$\begin{aligned}\sum_P \nu_P(x)d(P) &= \sum_{P\in S} \nu_P(x)d(P) - \sum_{P\in T} \nu_P(x^{-1})d(P)\\ &= d(Z) - d(Y) = |F : K(x)| - |F : K(x^{-1})| = 0,\end{aligned}$$

since $K(x) = K(x^{-1})$. □

11.10 Principal divisors

The degree theorem suggests that we define a further divisor associated with x.

DEFINITION The divisor $[x] = [x]_0 - [x]_\infty$ is called the *divisor of* x. Divisors $[f]$ of non-zero functions are called *principal* divisors.

Two divisors A and B for which $A - B$ is principal are said to be *equivalent* or to lie in the same *class*.

Principal divisors form a subgroup of the set of all divisors, because the divisor of x/y is the difference of the divisors of x and y. They all have degree 0.

PROPOSITION *If A and B are equivalent divisors, then $d(A) = d(B)$ and $l(A) = l(B)$.*

Proof Suppose that $A - B = [y]$. Then $d(A) = d(B) + d([y]) = d(B)$. Furthermore, the map $z \mapsto zy$ is a linear bijection from $L(A)$ to $L(B)$. Hence we also have $l(A) = l(B)$. □

11.11 Riemann's theorem and the genus

We shall now show that the inequality of Riemann proved for divisors of zeros in Proposition 11.8 holds for all divisors D. The proof given here follows Deuring (1973).

THEOREM Riemann's theorem *There exists an integer t such that the inequality $l(D) \geq d(D) - t$ holds for every divisor D. Any such t satisfies $t \geq -\kappa$.*

Proof We have already established the first statement if D is the divisor of zeros of a non-constant function. The idea of the proof is to use the inequality of Proposition 11.6 to compare D with the divisor of zeros of a non-constant element. So fix x non-constant with divisor of zeros $Z = [x]_0$ and let t be the constant found by Proposition 11.8. If $D \leq mZ$ for some m, then by Proposition 11.6, $l(D) - d(D) \geq l(mZ) - d(mZ) \geq -t$. So $l(D) \geq d(D) - t$.

We now embark on a detour. We shall first replace D by a positive divisor D' and then replace that divisor by an equivalent divisor D'' that satisfies the condition $D'' \leq mZ$. For the first stage of the detour we represent the divisor D as the difference between two positive divisors: $D = D' - D^-$ with $D', D^- \geq N$. Then $D \leq D'$, so by Proposition 11.6, $l(D) - d(D) \geq l(D') - d(D')$. Hence if the inequality holds for D' it holds for D as well.

For the second stage of the detour recall that D' is equivalent to $D' - [y]$ for any non-zero element y of F. It follows that $l(D' - [y]) = l(D')$ and $d(D' - [y]) = d(D')$. Hence if the inequality holds for $D' - [y]$ it holds for D'.

We now apply Proposition 11.6 a third time to find an element y such that $D'' = D' - [y] \leq mZ$ for some m. Since $D' \geq N$ we have $mZ \geq mZ - D'$. Hence

$$l(mZ - D') - d(mZ - D') \geq l(mZ) - d(mZ') \geq -t.$$

Therefore

$$l(mZ - D') \geq d(mZ - D') - t = md(Z) - d(D') - t.$$

By construction $d(Z) > 0$. Hence for sufficiently large m, $l(mZ - D') > 0$.

Taking $0 \neq y \in L(mZ - D')$, we have $\nu_P(y) + mZ(P) - D'(P) \geq 0$ or $mZ(P) \geq D'(P) - \nu_P(y)$ for all P. Therefore $D' - [y] \leq mZ$. Thus Riemann's inequality holds for $D'' = D' - [y]$. Then it also holds for D' and therefore for the original divisor D. That proves the main claim of the theorem.

If we take D to be the null divisor N, then $l(N) = \kappa$ and $d(N) = 0$, so $l(N) = d(N) + \kappa$. It follows that any t for which $l(D) \geq d(D) - t$ for all divisors must be at least $-\kappa$. □

DEFINITION Let $\gamma - \kappa$ be the minimal value of t for which Riemann's theorem holds. Then γ is called the *genus* of F over K. If we need to distinguish the base field we shall write γ_K.

The proof of Riemann's theorem shows that the genus is determined by any divisor of zeros.

COROLLARY *Let x be a non-constant element of F and let $Z = [x]_0$ be its divisor of zeros. Then γ is the minimal integer for which $l(mZ) \geq d(mZ) + \kappa - \gamma$ holds for all positive m.* □

11.12 Rational function fields

Genus 0 almost characterizes fields of rational functions, but if the field K is not algebraically closed there are exceptions as some examples will show.

THEOREM *Let F be a field of algebraic functions over K. Then the following three statements are equivalent.*
(i) *$F = K(x)$ for some $x \in F$.*
(ii) *F contains a place P of degree 1 such that $l(P) > 1$.*
(iii) *The genus γ is 0 and F contains a place of degree 1.*

Proof
(i) ⇒ (ii) If $F = K(x)$, then the divisor of zeros of $Z = [x]_0$ has support consisting of a single place P. In F_P/P every polynomial in x is mapped to its constant term. So every rational function is mapped to a quotient of elements of K. Thus $F_P/P = K$ and this place has degree 1.

It follows also that the field of constants $\bar{K} = K$ and so $L(N) = K$. As $1/x \in L(P)$ it follows that $l(P) > 1$.

(ii) ⇒ (iii) For any element $u \in L(P) \setminus L(N)$ we have

$$u^m \in L(mP) \setminus L((m-1)P).$$

Hence $l(mP) \geq l(m-1)P + 1$ for all m. Since $l(P) \geq 2$, it follows that $l(mP) \geq m + 1$ for all m. Choose $1/x \in L(P) \setminus L(N)$. Then $[x]_0 = P$ and the corollary of Riemann's theorem gives $\gamma = 0$.

(iii) ⇒ (i) Let P be a place of degree 1. Since $d(P) \geq \kappa$ it follows that $K = \bar{K}$. As the genus is zero Riemann's theorem implies that $l(P) \geq d(P) + 1 = 2$. Hence $L(P)$ contains an element $1/x \notin K$. Then P is the divisor of zeros of x. Hence $|F : K(x)| = d(P) = 1$. Thus $F = K(x)$. □

In the theorem the assumption that a place of degree 1 exists cannot be omitted. If the base field is algebraically closed then all places will be rational so the theorem holds. Otherwise a field of the form $L(x)$ where L is an algebraic extension of K has genus 0 but does not satisfy the proposition. That can occur in a slightly less obvious fashion as the first example below shows. If that were the only problem, it could be avoided by insisting that the field of constants is K. However, our second example shows that this assumption does not suffice and that the field may not be isomorphic to $L(x)$ for any algebraic extension.

EXAMPLES Let $K = \mathsf{R}$ and consider the plane 'curve' C: $y^2 + 1 = 0$ which has no real points. The function field $F = K(C)$ is just the field $\mathsf{C}(x)$ of rational functions over the complex numbers. The element x has a single zero of degree 2 corresponding to the conjugate points $(0, \pm\mathrm{i})$. We denote this place by P.

Then the functions 1, $1/x$, ..., $1/x^m$, i, i/x, ..., i/x^m all lie in $L(mP)$ and are linearly independent. Thus $l(mP) \geq d(mP) + 2 = d(mP) + \kappa$. Therefore by the corollary to Riemann's theorem the genus of F is 0.

Now we consider a more complicated example.

Again let $K = \mathsf{R}$ and consider the curve C: $x^2 + y^2 + 1 = 0$, which also has no real points, and let F be the function field $K(C)$. If σ were an isomorphism of F onto the rational function field $\mathsf{R}(t)$, then $\sigma(x) = \varphi(t)$ and $\sigma(y) = \psi(t)$ would be non-zero real rational functions satisfying $\varphi^2 + \psi^2 + 1 = 0$, but no such functions exist. So F is not a field of rational functions over R.

Nevertheless, the field of constants is R, as the following argument shows.

We can use the equation of C to eliminate $y^2 = -(x^2 + 1)$ from the polynomials in $K[C]$, so that every element of $K[C]$ has a unique representation in the form $f(x) + yg(x)$ where f and g are polynomials in x alone.

In our calculations we shall use the fact that the subring $\mathsf{R}[x]$ of F is a true ring of polynomials. If the field of constants of F were C, then there would be an element of F with square -1. We can write $(a+yb)^2/(c+yd)^2 = -1$ where a, b, c, d are polynomials in x. Multiplying this equation out, eliminating y^2, and gathering terms in y we obtain

$$a^2 + c^2 - (x^2 + 1)(b^2 + d^2) = -2y(ab + cd).$$

As y does not occur on the left hand side of the equation it follows that $ab = -cd$. Letting $s(x) = c/a = -b/d$, we get

$$a^2(1 + s^2) - (x^2 + 1)(1 + s^2)d^2 = 0.$$

Since s is a real rational function in x, its square is not -1 and we can cancel the terms $1 + s^2$. Then it follows that $1 + x^2 = (a/d)^2$. This equation has no solution in $\mathsf{R}(x)$ and so F has no element with square -1. That establishes the claim that the field of constants of F is $\mathsf{R} = K$. So F is not a field of rational functions over any field.

The divisor of zeros of x is a single place of degree 2 corresponding to the points $(0, \pm\mathsf{i})$. Hence the divisor Z of poles of x also has degree 2. By Proposition 11.7 the divisor of poles of x^2+1 is $2Z$. The equation $x^2+1 = -y^2$ now implies that Z is also the divisor of poles of y.

Therefore $L(mZ)$ contains the elements $1, x, \ldots, x^m, y, yx, \ldots, yx^{m-1}$, which are linearly independent. Hence

$$l(mZ) \geq 2m + 1 = d(mZ) + 1.$$

It follows from the corollary to Riemann's theorem that F has genus 0.

11.13 Exercises

1. Let C: $f(x, y) = 0$ be a curve defined over K and let (α, β) be a simple point on C with local ring V. Denote the maximal ideal of V by P also, and show that $V/P = K[\alpha, \beta]$.
2. Prove that the set $L(D, S)$ forms a vector space over K.

3. Prove Corollary 11.1: If F has a rational place, then the constant degree κ is 1.

4. Let $U \subseteq V \subseteq W$ be three vector spaces over a field. Prove that the relative dimensions satisfy the formula $\dim W/U = \dim W/V + \dim V/U$.

Use this to conclude the proof of Proposition 11.3 using induction on $d(B) - d(A)$.

Hint: First show that if $d(B) > d(A)$ then there exists a place Q such that $A < A + Q \leq B$.

5. Let x be a non-constant element of the function field F and let $u = f(x)/g(x)$ where f and g are polynomials in x. Show that

$$[u] = [f]_0 - [g]_0 - (\deg f - \deg g)[x]_\infty.$$

6. Continuing with the assumptions of the previous exercise, assume that f and g have no common factors. Show that then $[f]_0$ and $[g]_0$ have disjoint support.

Deduce that if $\deg f \geq \deg g$, and f and g have no common factors, then the divisors of zeros of $u = f/g$ and f are the same. Show further that the degree $|K(x) : K(u)|$ is equal to the larger of $\deg f$ and $\deg g$.

7. Let D be a divisor with $d(D) \leq 0$. Show that $l(D) = 0$ unless D is principal, in which case $d(D) = 0$ and $l(D) = \kappa$.

8. Let $F = K(x)$ be a field of rational functions and let P be any place of F. Show that $l(P) = d(P) + 1$. Exhibit a basis of $l(P)$ where P corresponds to the irreducible polynomial $g(x)$.

12

REPARTITIONS AND DIFFERENTIALS

In order to prove the full Riemann–Roch theorem it is necessary to introduce the concept of a differential. Originally this was done analytically, but that limits the applicability of the theory to real and complex curves. In the 1920s André Weil found a purely algebraic way of defining differentials. Apart from extending the theory to arbitrary fields, this had the additional advantage that it simplified the proofs of the main theorems. However, the definitions were both abstract and difficult to link to the usual idea of a differential.

In this chapter I shall present a slightly modified version of Weil's concepts together with some sections motivating the definitions, which will, I hope, make them less daunting.

12.1 Classical integration

We shall base the definitions we make on the classical theory of complex integration and ignore differentiation even though it provides the name 'differential'. In classical complex analysis the integral

$$\int_{\Gamma} f dz$$

is an operator that produces a number from three inputs:

1. a path Γ, which in the most important cases is closed and simple,
2. a function f with no poles on Γ, and
3. a differential dz.

The differential is the most mysterious of these three inputs, because its properties are usually only implicitly defined. Note, however, that it may vary and that it then obeys the formula

$$dw = \frac{dw}{dz} dz.$$

For our purposes dw/dz is just another function so that differentials form a one-dimensional vector space over the space of functions. The differential dz can be viewed as mapping path-function pairs (Γ, f) to the base field C by taking (Γ, f) to $\int_{\Gamma} f dz$. With respect to f the action is linear and it can also be regarded as linear on the Γ component if we introduce a suitable addition of paths. Our aim is to introduce an algebraic object for an arbitrary function field that exhibits the properties of path-function pairs. Then we shall be able to define differentials as special linear maps on these objects.

12.2 Repartitions

To remove the topological aspect of the path Γ in the classical theory we use the classical residue theorem:

$$\int_{\Gamma} f dz = \sum_{P \in \Gamma^{\circ}} \operatorname{res}(f dz),$$

where Γ° is the interior of the closed simple path Γ (often you will just find $\operatorname{res}(f)$, which assumes that the differential is dz). Classical theory evaluates the sum on the right and shows that it is finite. We shall return to that topic later. For the moment observe that the residue theorem allows us to replace Γ by Γ°, which can be treated as a collection of points. In this way we can replace the pair (Γ, f) by a 'selection mapping' associating f with all points in Γ° and 0 with all other points. In order to allow linear operations, we extend our consideration to the set of all finite sums of such selections and make the following definition[1].

DEFINITION A *repartition* of F/K is a map Γ assigning to each place of F/K an element of F such that Γ has only finitely many distinct values. We write Γ_P for the image of P under Γ and define $\nu_P(\Gamma) = \nu_P(\Gamma_P)$.

If S is any set of places and $f \in F$ we denote by f/S the repartition that maps P to f if $P \in S$ and to 0 otherwise. This is the object that corresponds to a path-function pair (Γ, f), with S playing the role of the interior Γ° of Γ. If S is the set of all places a repartition f/S is called *constant*. Constant repartitions are identified with elements $f \in F$ and denoted by the same symbol f. For general S the complementary partition $f - f/S$ which maps P to f if $P \notin S$ and to 0 otherwise is denoted by $f \setminus S$.

Note that Γ_P is an element of F. So a repartition can be viewed as first partitioning the set of all places into finitely many subsets $S_1, \ldots, S_r$ and then choosing a function f_i for each of these subsets. Such a repartition is a sum $f_1/S_1 + \cdots + f_r/S_r$.

Our first aim is to generalize the theory of L-spaces of the previous chapter to repartitions. To that end we make some further definitions.

DEFINITION The set of all repartitions is denoted by X. We define the sum of two repartitions Γ and Δ by $(\Gamma + \Delta)_P = \Gamma_P + \Delta_P$ for all places. For any function $f \in F$ we define the repartition $f\Gamma$ by $(f\Gamma)_P = f\Gamma_P$ for all places.

With these operations X becomes a vector space over F and hence also over K. It is generated over K by the partitions of the form f/S.

In analogy to $L(D)$ we define $X(D)$ for a divisor to be the set of repartitions Γ such that

$$\nu_P(\Gamma) + D(P) \geq 0$$

for all places P. It is called the *space of repartitions over D*.

[1]In the literature repartitions or *adèles* are defined as a space which strictly contains our space.

We shall consider these sets as vector spaces over K, leaving the straightforward verification of the vector space properties to the exercises.

12.3 Relative dimensions

We first prove an analogue of Proposition 11.3.

PROPOSITION *Let A and B be divisors with $A \leq B$. Then $X(A) \subseteq X(B)$ and* $\dim X(B)/X(A) = d(B) - d(A)$.

Proof The inclusion follows directly from the definition.

Let S be a finite set of places including the supports of A and B. We shall establish an isomorphism from $L(B,S)/L(A,S)$ to $X(B)/X(A)$. Then the formula follows from Proposition 11.3.

The map we shall use is simply this. Take $u \in L(B,S)$ to u/S. It is clear that $u/S \in X(B)$ and that $u \in L(A,S)$ if and only if $u/S \in X(A)$. Hence our map determines an injection φ from $L(B,S)/L(A,S)$ to $X(B)/X(A)$.

To show that φ is surjective, choose $\Delta \in X(B)$. By the approximation theorem there exists $w \in F$ such that $\nu_P(w - \Delta_P) \geq -A(P)$ for all P in S. As $\nu_P(\Delta) \geq -B(P)$ it follows that $\nu_P(w) \geq -B(P)$ for all $P \in S$. Hence $w \in L(B,S)$. Now for $P \in S$ we have $\nu_P(w/S - \Delta) \geq -A(P)$ and for $P \notin S$ we have $(w/S - \Delta)_P = -\Delta_P$ and $\nu_P(\Delta) \geq 0$. Hence $w/S - \Delta \in X(A)$. Therefore φ is an isomorphism. □

12.4 Differentials

Now that we have defined repartitions as the objects on which differentials act, we must ask what special properties should be used to distinguish differentials from arbitrary linear maps. There are two particularly striking special cases of the classical residue theorem.

THEOREM A *The sum of the residues of $f\,dz$ is zero.* □

THEOREM B Cauchy's theorem *If Γ is a path such that Γ° contains no poles of $f\,dz$, then*

$$\int_\Gamma f\,dz = 0.$$ □

In the context of our abstract definition, Theorem A applies to a constant repartition f. It says that the linear map corresponding to a differential dz should take such a repartition to 0. Extending Cauchy's theorem requires us to look at the poles of $f\,dz$.

EXAMPLE Consider the field $F = \mathsf{C}$ and the 'differential' $f\,dz$. The finite poles of this are just the poles of f. To see what happens at infinity we introduce the variable $w = 1/z$. Then $z = 1/w$ and so $dz = -dw/w^2$. Thus it makes sense to put $\nu_\infty(f\,dz) = \nu_\infty(f) - 2$. In particular $\nu_P(dz) = 0$ for all finite places and $\nu_\infty(dz) = -2$. For general f, $\nu_P(f\,dz) = \nu_P(f) + \nu_P(dz)$. This formula also allows us to calculate $\nu_P(df)$ by replacing df by $f'dz$.

From this discussion we can restate Cauchy's theorem as

THEOREM B′ *If $\nu_P(f) + \nu_P(dz) \geq 0$ for all $P \in \Gamma^\circ$, then $\int_\Gamma f\,dz = 0$.* □

These two results motivate the following definition.

DEFINITION A K-linear map ω from X to K is called a *differential* if $\omega(f) = 0$ for all constant repartitions and there exists a divisor D such that $\omega(\Gamma) = 0$ for all repartitions $\Gamma \in X(D)$.

For convenience we introduce the notation $Y(D)$ for the space of all repartitions Γ of the form $\Delta + y$ with $\Delta \in X(D)$ and $y \in F$ and call this the *extended space of repartitions over D*. Then the condition can be reformulated as $\omega(\Gamma) = 0$ for all $\Gamma \in Y(D)$ for some divisor D.

12.5 Relative dimensions again

We can calculate the dimension of $Y(B)/Y(A)$ in a manner similar to the calculation for $X(B)/X(A)$.

PROPOSITION *Let A and B be divisors and $A \leq B$. Then $Y(A) \subseteq Y(B)$ and* $\dim Y(B)/Y(A) = d(B) - d(A) - (l(B) - l(A))$.

Proof Before starting the proof recall two standard results on vector spaces (considered as abelian groups). The first of these is the isomorphism theorem (see Paragraph 11.6); the second is the homomorphism theorem.

THEOREM Isomorphism theorem *Let $U, V \subseteq W$ be vector spaces over a field. Then*

$$U/U \cap V \cong (U + V)/V. \qquad \square$$

THEOREM Homomorphism theorem *Let $U \subseteq V \subseteq W$ be three vector spaces over a field. Then*

$$W/V \cong (W/U)/(V/U). \qquad \square$$

Since, by its definition, $Y(B) = X(B) + Y(A)$ it follows by the isomorphism theorem that $Y(B)/Y(A) = X(B)/(X(B) \cap Y(A))$. To evaluate the denominator, notice that for $\Gamma \in X(A)$ and $y \in F$, the statement $\Gamma + y \in X(B)$ holds if and only if $y \in X(B) \cap F$ and the definitions directly yield $X(B) \cap F = L(B)$. So $X(B) \cap Y(A) = X(A) + L(B)$. Thus

$$\begin{aligned} Y(B)/Y(A) &\cong X(B)/(X(B) \cap Y(A)) \\ &= X(B)/(X(A) + L(B)) \\ &\cong \frac{X(B)/X(A)}{(X(A) + L(B))/X(A)} \\ &\cong \frac{X(B)/X(A)}{L(B)/(L(B) \cap X(A))} \\ &= \frac{X(B)/X(A)}{L(B)/L(A)}. \end{aligned}$$

Evaluating the dimensions of the last numerator and denominator separately we obtain $\dim Y(B)/Y(A) = (d(B) - d(A)) - (l(B) - l(A))$. $\square$

We can use Riemann's theorem to estimate $d(B) - l(B)$ and thereby find an estimate for $\dim Y(B)/Y(A)$ that is independent of B.

COROLLARY *If $A \leq B$, then $\dim Y(B)/Y(A) \leq l(A) - d(A) + \gamma - \kappa$, with equality if $d(B) - l(B) = \gamma - \kappa$.* □

12.6 The index of a divisor

As we shall prove in the next paragraph, Corollary 12.5 implies that $\dim X/Y(D)$ is finite for all divisors. Indeed it turns out to be the missing term in Riemann's theorem. That is the reason for the following definition.

DEFINITION The *index* of a divisor D is $j(D) = \dim X/Y(D)$.
As both X and $Y(D)$ are vector spaces over $\bar{K}$ it follows that $j(D)$ is a multiple of κ.

EXAMPLE Let $F = K(x)$ be a field of rational functions and let P be the infinite place corresponding to $1/x$. We shall calculate $j(-nP)$ for some values of n.

In the case $n = 1$ we have $Y(-P) = X$. This can be shown by proving that $y/S \in Y(-P)$ for all $y \in F$ and all sets of places S. Since $y \in Y(D)$ for all divisors we have in particular $y \in Y(-P)$. So y/S lies in $Y(-P)$ if and only if $y \setminus S = y - y/S$ lies in $Y(-P)$. Since $y \setminus S = y/T$, where T is the set of places not in S, we may assume that $P \notin S$. We argue by induction on the number of poles of y in S. If there are none, then $y/S \in X(-P)$.

Let Q be a pole of y in S corresponding to the irreducible polynomial t and suppose the order of y at Q is $-r$. Then by Corollary 10.5,

$$t^r y \equiv f_r \pmod{t},$$

for some unit $f_r \in F_Q$. So $y - f_r/t^r$ has order at least $-r+1$ at Q. The element f_r is only determined modulo Q so we may take f_r to be a polynomial of degree less than $\deg t$. Repeating this argument, we find polynomials f_i of degree less than $\deg t$ for $i = 1, \ldots, r$ such that

$$y - \sum_{i=r}^{1} \frac{f_i}{t^i} = y - \frac{\sum_{i=1}^{r} f_i t^{r-i}}{t^r}$$

has order ≥ 0 at Q.

We rewrite the right hand side as $y - s$. Observe that the only pole of s is at Q because t is the only irreducible polynomial dividing the denominator and the degree of the numerator is less than the degree of the denominator, so that P is not a pole (indeed that shows that s has order at least 1 at P). Hence $s \setminus S \in Y(-P)$ and therefore $s/S \in X(-P)$. It follows that $y/S \in Y(-P)$. Thus $Y(-P) = X$ and so $j(-P) = 0$. As the index is a decreasing function of the set of divisors, it follows that $j(D) = 0$ for all positive divisors.

For the case $n = 2$ the situation changes. The repartition x^{-1}/P forms a basis of X modulo $Y(-2P)$ and thus $j(-2P) = 1$. To prove this we first show that x^{-1}/P does not lie in $Y(-2P)$. For suppose that $x^{-1}/P = \Gamma + z$, where $\Gamma \in X(-2P)$ and $z \in K(x)$. Then $\nu_P(z) = 1$ because $\nu_P(x^{-1}) = 1$ and by definition $\nu_P(\Gamma) \geq 2$. It follows that $\nu_Q(z) < 0$ for some $Q \neq P$, while $\nu_Q(\Gamma) \geq 0$. That implies that $\nu_Q(\Gamma + z) < 0$ contradicting the fact that $(x^{-1}/P)_Q = 0$.

To show that x^{-1}/P forms a basis of X modulo $Y(-2P)$ we exploit the fact that $X = Y(-P)$. Thus an arbitrary element of X can be written in the form $\Delta + y$ with $\Delta \in X(-P)$ and hence $\nu_P(\Delta) \geq 1$. If $\nu_P(\Delta) \geq 2$, then $\Delta \in X(-2P)$ and so $\Delta + y \in Y(-2P)$. So assume that $\nu_P(\Delta) = 1$. Then since x^{-1} is a local parameter for P it follows from Corollary 10.5 that $\nu_P(\Delta_P - ax^{-1}) \geq 2$ for some unit $a \in F_P$. The element a is only determined modulo P and P is a rational place. Hence we can take $a \in K$. That implies that $\Delta - (ax^{-1})/P \in X(-2P)$ and so $\Delta + y - ax^{-1}/P \in Y(-2P)$. Therefore $\Delta + y \equiv ax^{-1}/P$ modulo $Y(-2P)$.

12.7 The weak Riemann–Roch theorem

I shall now show, as promised, that the index of D is the missing term in Riemann's theorem. I call this the weak Riemann–Roch theorem, since, as you have seen above, the index is far from easy to calculate. In the strong form we shall find that the index can be calculated as the rank $l(W - D)$ for a suitably chosen divisor W.

THEOREM Weak Riemann–Roch theorem *For every divisor D,*

$$l(D) = d(D) + j(D) + \kappa - \gamma.$$

Proof By Riemann's theorem there exists a divisor C with $d(C) - l(C) = \gamma - \kappa$, and then $d(B) - l(B) = \gamma - \kappa$ holds for all divisors $B \geq C$. Suppose that $B \geq C \cup D$, that is, $B(P) \geq \max\{C(P), D(P)\}$ for all P. Then by Corollary 12.5,

$$\dim X/Y(D) \geq \dim Y(B)/Y(D) = l(D) - d(D) + \gamma - \kappa.$$

Hence $j(D) \geq l(D) - d(D) + \gamma - \kappa$.

On the other hand let $k = l(D) - d(D) + \gamma - \kappa$ and suppose $\Gamma_0, \ldots, \Gamma_k \in X$. Since Γ_{iP} takes only finitely many values, there is a positive divisor A_i such that $\Gamma_i \in X(A_i)$. Then letting $A(P) = \min\{A_0(P), \ldots, A_n(P)\}$ for all P it follows that $\Gamma_i \in X(A) \subseteq Y(A)$ for $i = 0, \ldots, k$. We can apply the first part of the proof to $B = A \cup C \cup D$. Since $\dim Y(B)/Y(D) = k$, it follows that $\Gamma_0, \ldots, \Gamma_k$ are linearly dependent in $Y(A \cup C \cup D)/Y(D)$. Then they are also linearly dependent in $X/Y(D)$ and hence $j(D) = \dim X/Y(D) \leq k$.

The two inequalities together give $j(D) = k$, proving the theorem. □

12.8 Differentials over D

In order to investigate the index we make a similar definition to those we have already made for functions and repartitions.

DEFINITION For a given divisor D the space of differentials that are zero on $X(D)$ (and hence automatically on $Y(D)$) is denoted by $J(D)$ and called the *space of differentials over* D. $J(D)$ can also be considered as the set of all K-linear maps from $X/Y(D)$ to K.

The main reason for introducing these spaces is that the last two statements of the following proposition hold.

PROPOSITION (a) *The index satisfies* $j(D) = \dim_K J(D)$.
(b) *If* $A \leq B$, *then* $J(B) \subseteq J(A)$.
(c) *If* $\omega \in J(A)$ *and* $x \in F$ *then the map* ωx *defined by* $\omega x(\Gamma) = \omega(x\Gamma)$ *is a differential over* $A + [x]$.
(d) *If* $0 \neq \omega \in J(A)$ *then the map* χ_ω: $x \mapsto \omega x$ *embeds* $L(A - B)$ *as a subspace of* $J(B)$.
(e) *If* $j(A) \neq 0$, *then* $l(A - B) \leq j(B)$ *for any* B.

Proof (a) $J(D)$ is just the K-dual of $X/Y(D)$. So by standard linear algebra $\dim J(D) = \dim X/Y(D) = j(D)$ (see Exercise 4).

(b) If $A \leq B$, then $Y(A) \subseteq Y(B)$. Thus if ω is zero on $Y(B)$ it is zero on $Y(A)$ *a fortiori*.

(c) This follows because if $\Gamma \in Y(A+[x])$, then $\Gamma = \Delta + y$ with $\Delta \in X(A+[x])$ and $y \in F$. Thus $\nu_P(\Delta)+\nu_P(x)+A(P) \geq 0$. Hence $x\Gamma = x\Delta + xy$ where $xy \in F$ and $x\Delta \in X(A)$. Therefore $\omega(x\Delta) = 0$ and hence also $\omega(x\Gamma) = \omega x(\Gamma) = 0$. Thus ωx is zero on $Y(A + [x])$ and so it is a differential.

(d) For $\Gamma \in X(B)$ and $x \in L(A - B)$, we have $\nu_P(\Gamma) + B(P) \geq 0$ and $\nu_P(x) + A(P) - B(P) \geq 0$ and so $x\Gamma \in X(A)$. Hence for $\Gamma \in Y(B)$ and $x \in L(A - B)$ we have $x\Gamma \in Y(A)$. It follows that $\omega x \in J(B)$ for $x \in L(A - B)$.

In order to show that this is an embedding, we must show that for $x \neq 0$, $\omega x \neq 0$. By assumption $\omega(\Delta) \neq 0$ for some $\Delta \in X$. Then $\omega x(x^{-1}\Delta) \neq 0$ and so ωx is non-zero. Since the mapping $x \mapsto \omega x$ is linear, it follows that it is an embedding.

(e) This follows directly from part (d). □

12.9 Properties of the index

The next proposition assembles some useful properties of the index.

PROPOSITION (a) *If* $A \leq B$, *then* $j(A) \geq j(B)$.
(b) *For the zero divisor* N, *the index* $j(N)$ *is* γ.
(c) *If* D *is a positive divisor with* $d(D) > \kappa - \gamma$, *then* $j(-D) > 0$. *So every function field has non-zero differentials.*
(d) *If* $j(D) > 0$, *then* $l(D) \leq \gamma$.
(e) *For any divisor* D *the index* $j(D)$ *is at most* $\max\{0, 2\gamma - \kappa - d(D)\}$. *In particular* $j(D) = 0$ *and* $Y(D) = X$ *for any divisor* D *with* $d(D) \geq 2\gamma - \kappa$.

Proof (a) This follows directly from Proposition 12.8a and b.

(b) We have $d(N) = 0$ and $L(N) = \bar{K}$, whence $l(N) = \kappa$. So by the weak Riemann–Roch theorem, $j(N) = l(N) - d(N) + \gamma - \kappa = \gamma$.

(c) Since $-D < N$, we have $l(-D) = 0$. Hence $0 = -d(D) + \kappa - \gamma + j(-D)$. The hypothesis on D ensures that $d(D) + \kappa - \gamma + j(-D) < j(-D)$.

(d) Noting that $D = D - N$, we see that by Proposition 12.8e, $l(D) \leq j(N) = \gamma$.

(e) If $j(D) = 0$ the statement is true. So assume that $j(D) \neq 0$. Then by part (d), $l(D) \leq \gamma$. Hence by the theorem of Riemann–Roch:

$$\gamma \geq l(D) = d(D) + j(D) + \kappa - \gamma.$$

Thus $j(D) \leq 2\gamma - \kappa - d(D)$ as claimed. □

12.10 The divisor of a differential

Although we allow the divisor used in the definition of a differential to be arbitrary, the maximal possible divisors for non-zero divisors are tightly constrained. The existence of a non-zero differential has been established in Proposition 12.9c.

PROPOSITION *Let ω be a non-zero differential. Then there is a unique maximal divisor D such that ω is zero on $Y(D)$.*

DEFINITION This divisor is called *the divisor of* ω and denoted by $[\omega]$. The *order* of ω at P denoted by $\nu_P(\omega)$ is the value $[\omega](P)$.

Proof We are given that ω is zero on $Y(D)$ for some D but not zero on X. By Proposition 12.9e every divisor D with ω zero on $Y(D)$ has $d(D) < 2\gamma - \kappa$. Thus there certainly exist maximal divisors D with the property that ω is zero on $Y(D)$. The task is to show that there is only one.

This will follow if we show that ω zero on $Y(A)$ and also on $Y(B)$ implies that ω is zero on $Y(A \cup B)$. By hypothesis, ω is zero on F and linear. Hence ω is zero on $Y(A)$ if and only if it is zero on $X(A)$. Suppose ω is zero on both $X(A)$ and $X(B)$ and let $\Gamma \in X(A \cup B)$. Define Δ and Λ as follows:

$$\begin{array}{lll} \Delta_P = \Gamma_P & \Lambda_P = 0 & \text{if } A(P) \geq B(P), \\ \Delta_P = 0 & \Lambda_P = \Gamma_P & \text{otherwise.} \end{array}$$

Then $\Delta \in X(A)$, $\Lambda \in X(B)$, and $\Gamma = \Delta + \Lambda$. By hypothesis $\omega(\Delta) = \omega(\Lambda) = 0$. So $\omega(\Gamma) = 0$.

Suppose now that A and B are distinct with ω zero on $Y(A)$ and $Y(B)$. Then ω is zero on $X(A \cup B)$ which cannot be equal to both A and B. So A and B are not both maximal. Thus there is only one maximal divisor D as claimed.

□

In the course of the proof we found that in searching for the divisor of ω we could replace $Y(A)$ by $X(A)$ and so established the first of the following corollaries. The multiplication ωx in the second corollary is the one defined in Proposition 12.8c: $\omega x(\Gamma) = \omega(x\Gamma)$ for all repartitions Γ.

COROLLARIES (a) *For a differential* ω *and a divisor* A, ω *is zero on* $X(A)$ *if and only if* $A \leq [\omega]$.
(b) *For a differential* ω *and an element* $x \in F$, *the divisor of* ωx *satisfies* $[\omega x] = [\omega] + [x]$.

Proofs Corollary (a) has already been proved. We use it to prove Corollary (b). If $\Gamma \in X([\omega]+[x])$ then $\nu_P(\Gamma) + \nu_P(\omega) + \nu_P(x) \geq 0$ for all places P. Therefore $\nu_P(x\Gamma) + \nu_P(\omega) \geq 0$ and hence $x\Gamma \in X([\omega])$. Thus $\omega(x\Gamma) = 0$ and so $\Gamma \in X([\omega x])$. That shows that $[\omega] + [x] \leq [\omega x]$. Writing $\omega = \omega x x^{-1}$ we obtain $[\omega x] + [x^{-1}] \leq [\omega]$. As $[x^{-1}] = -[x]$ equality follows. □

EXAMPLE Let $F = K(x)$, the field of rational functions. We shall calculate the divisor of dx. The first question we must answer is 'what is the differential dx?' In Example 12.4 we found that the divisor of dx should be $-2P$, where P is the place at infinity. Applying the Riemann–Roch theorem to this divisor and substituting known values for $l(-2P) = 0$, $d(-2P) = 0$, $\gamma = 0$, and $\kappa = 1$ we obtain

$$0 = l(-2P) = d(-2P) + j(-2P) + \kappa - \gamma = j(-2P) - 1.$$

So up to constant multiples there is only one differential with divisor $-2P$. In Example 12.6 we found that $X/Y(-2P)$ is one dimensional and generated by x^{-1}/P. Thus we can legitimately define dx by setting $dx(x^{-1}/P) = 1$ and $dx(\Delta) = 0$ for $\Delta \in Y(-2P)$. This corresponds to the classical complex integral formula

$$\frac{1}{2\pi \mathrm{i}} \int_\Gamma \frac{1}{z} dz = 1$$

for a circle Γ around the origin, except that we have removed the constant $2\pi\mathrm{i}$.

We check that the divisor of dx is indeed $-2P$. As $dx(\Gamma) = a$ if and only if $\Gamma - ax^{-1}/P \in Y(-2P)$ it follows that $dx(\Gamma) = 0$ if and only if $\Gamma \in Y(-2P)$. We need therefore only show that $Y(D) = X$ for $D > -2P$. We use the estimate of Proposition 12.9e. By hypothesis $d(D) \geq -1$, while for $K(x)$ we have $\kappa = 1$ and $g = 0$. Hence $2\gamma - \kappa - d(D) \leq 0$. Therefore by the estimate $j(D) = 0$, and thus $Y(D) = X$.

12.11 Differentials as multiples

In this paragraph we shall show that the multiplication ωx that was defined in Proposition 12.8c makes the set of differentials into a one-dimensional vector space over F. That is a partial generalization of the familiar formula $dw = (dw/dz)dz$ from the classical case.

THEOREM *Let* ω *be a non-zero differential. To any differential* σ *there exists a unique* $x \in F$ *such that* $\omega x = \sigma$.

Proof (1) We establish uniqueness first. Suppose that $\omega x = \omega y$ for two elements $x, y \in F$. We may assume that one of the elements, say y, is non-zero. It follows that $\omega = \omega x/y$. Therefore by Corollary 12.10b, $[\omega]$ and $[\omega] + [x/y]$ must be the same divisor. Hence $[x/y] = N$, the null divisor, and so x/y is a constant.

We shall show that the only constant z with $\omega z = \omega$ is $z = 1$. For, since $z \in \bar{K}$, it is algebraic over K. Let $t(\dot{x}) = \sum a_i \dot{x}^i$ be the minimum polynomial of z over K. Then $t(z) = 0$ and so

$$\omega \sum a_i z^i = 0,$$

where the right hand side denotes the zero divisor. The fact that $\omega z = \omega$ implies that $\omega z^i = \omega$ and so $\omega \sum a_i = 0$. If $\sum a_i \neq 0$ then $\omega = 0(\sum a_i)^{-1}$ which contradicts the assumption that $\omega \neq 0$. Hence $\sum a_i = 0$. In other words 1 is a root of $t(\dot{x})$ and so $(\dot{x} - 1)$ divides $t(x)$ which is irreducible. Hence $t(\dot{x}) = \dot{x} - 1$ and therefore $z = 1$.

That proves that $x = y$ and establishes that there is at most one function x with $\omega x = \sigma$.

(2) We now prove the existence of x for arbitrary σ. If $\sigma = 0$ we can take $x = 0$, so assume that $\sigma \neq 0$. Let $A = [\omega]$ and $B = [\sigma]$. Let D be a positive divisor that will be required to have a large degree m. We will note the conditions that m must satisfy as we require them. All will be lower bounds.

By Proposition 12.8d $L(A+D)$ and $L(B+D)$ are embedded in $J(-D)$. We require m to be sufficiently large that both $l(A+D) \neq 0$ and $l(B+D) \neq 0$. By Riemann's theorem that will hold, provided that $m > \gamma - \kappa - d$ where $d = \min\{d(A), d(B)\}$.

By the weak Riemann–Roch theorem, we have $j(-D) = l(-D) - d(-D) + \gamma - \kappa$ and for positive D, $l(-D) = 0$. Hence $j(-D) = m + \gamma - \kappa$. On the other hand applying Riemann's theorem to $A + D$ and $B + D$ gives

$$l(A+D) + l(B+D) \geq d(A) + d(B) + 2m + 2\kappa - 2\gamma.$$

That will imply that $l(A+D)+l(B+D) \geq m+\gamma$ provided that $m \geq 3\gamma - 2\kappa - d(A) - d(B)$.

(3) In that case $l(A+D)+l(B+D) > j(-D)$, and so the images of $L(A+D)$ and $L(B+D)$ must have a non-zero differential $\tau \in J(-D)$ in common. Suppose that τ is the image of $x \in L(A+D)$ and $y \in L(B+D)$. From the mapping that embeds these spaces in $J(-D)$ we have $\tau = \omega x = \sigma y$. Then $y \neq 0$ and $\sigma = \omega x / y$. □

DEFINITION A divisor is called *canonical* if it is the divisor of a non-zero differential.

COROLLARIES (a) *If ω is a non-zero differential, then $j([\omega]) = \kappa$.*
(b) *If A and B are canonical divisors, then $B = A + [x]$ for some non-zero function x.*

Proofs (a) Let $\omega x \in J([\omega])$. By Corollary 12.10b the divisor of ωx is $[\omega] + [x]$. Hence $[\omega] + [x] \geq [\omega]$ and so $[x] \geq N$. That is only possible if x is a constant. Therefore $J([\omega]) = \bar{K}\omega$ and hence $j([\omega]) = \kappa$.

(b) If A is the divisor of ω and B is the divisor of σ, then $\sigma = \omega x$ for some non-zero x and therefore $B = A + [x]$. □

EXAMPLE The proposition shows that every divisor of the field of rational functions $K(x)$ has the form fdx for some function $f \in K(x)$ and that $fdx = gdx$ if and only if $f = g$. It is tempting to identify differentials with functions, but they are different objects and the association between them depends on a choice of 'base' differential and there is no natural way to fix this choice.

It is possible to define $df = f'dx$ where f' is the derivative of f. However not every element of $K(x)$ has the form f' and so not every differential of $K(x)$ has the form df.

12.12 The strong Riemann–Roch theorem

Theorem 12.11 is the key that produces the strong version of the Riemann–Roch theorem.

THEOREM Strong Riemann–Roch theorem *Let W be a canonical divisor; then $j(D) = l(W - D)$ for any divisor D. Hence*

$$l(D) = d(D) + \kappa - g + l(W - D).$$

Proof By definition W is the divisor of a non-zero differential, say ω. We shall show that the map $x \mapsto \omega x$ induces an isomorphism between $L(D)$ and $J(W-D)$. If $x \in L(D)$, then $\nu_P(x) \geq -D(P)$ for all places P. Hence $\nu_P(\omega x) \geq \nu_P(\omega) - D(P) = (W-D)(P)$ and $\omega x \in J(W-D)$. Conversely if $\sigma \in J(W-D)$, then $\sigma = \omega x$ for some $x \in F$, and $\nu_P(x) = \nu_P(\sigma) - \nu_P(\omega) \geq -D(P)$ for all places P. Hence $x \in L(D)$.

Replacing D by $W - D$, it follows that $l(W - D) = j(D)$ and hence by the weak Riemann–Roch theorem

$$l(D) = d(D) + \kappa - \gamma + j(D) = d(D) + \kappa - \gamma + l(W - D). \qquad \square$$

COROLLARIES (a) *For a canonical divisor W, we have $l(W) = \gamma$ and $d(W) = 2\gamma - 2\kappa$.*
(b) *For any divisor D with $d(D) > 2\gamma - 2\kappa$ Riemann's inequality is an equation:*

$$l(D) = d(D) + \kappa - \gamma.$$

Proofs (a) By the theorem $l(W) = j(N)$, where N is the null divisor. By Proposition 12.9b, $j(N) = \gamma$ and hence $l(W) = \gamma$. From Corollary 12.11 we know that $j(W) = \kappa$. Hence by the theorem,

$$d(W) = l(W) - \kappa + \gamma - j(W) = 2\gamma - 2\kappa.$$

(b) The hypothesis implies that $d(W - D) < 0$ for a canonical divisor W. Therefore $l(W - D) = 0$ and so

$$l(D) = d(D) + \kappa - \gamma. \qquad \square$$

12.13 The strong approximation theorem

One major consequence of the Riemann–Roch theorem is the following significant strengthening of the approximation theorem and its corollary.

THEOREM Strong approximation theorem *Let $P_1, \ldots, P_h$ be distinct places with valuation functions ν_i and let Q be a further place. Suppose we are given $v_k \in F$ and $m_k \in \mathsf{Z}$ for each place P_k. Then there exists $x \in F$ such that*

$$\nu_k(x - v_k) \geq m_k$$

for all $k = 1, \ldots, h$ and $\nu_P(x) \geq 0$ for all other places P except Q. Furthermore, if m satisfies

$$md(Q) - \sum_{k=1}^{h} m_k d(P_k) > 2\gamma - 2\kappa,$$

then x can be found such that also $\nu_Q(x) \geq -m$.

Proof Consider the repartition Γ with

$$\Gamma_P = \begin{cases} v_k & \text{if } P = P_k, \\ 0 & \text{otherwise.} \end{cases}$$

Define $D = mQ - \sum(m_k)P_k$ choosing m sufficiently large, so that $d(D) > 2\gamma - 2\kappa$. Corollary (b) of the strong Riemann-Roch theorem then implies that $j(D) = 0$ and hence $Y(D) = X$. So there exists $x \in F$ such that $\Gamma - x \in X(D)$. Thus

$$\nu_k(\Gamma - x) = \nu_k(v_k - x) \geq m_k,$$

and for any other $P \neq Q$,

$$\nu_P(\Gamma - x) = \nu_P(x) \geq 0.$$

Finally for Q

$$\nu_Q(\Gamma - x) = \nu_Q(x) \geq -m. \qquad \square$$

COROLLARY *For any sequence $n_1, \ldots, n_h$ there exists an element x such that $\nu_k(x) = n_k$ for all $k = 1, \ldots, h$ and $\nu_P(x) \geq 0$ for all other places P except Q. Furthermore, if $nd(Q) - \sum(n_k + 1)d(P_k) > 2\gamma - 2\kappa$, then x can be found such that also $\nu_Q(x) \geq -n$.*

Proof This follows from the strong approximation theorem exactly as the original corollary did from the Artin–Wharples approximation theorem. For each k let $\nu_k(t_k) = 1$. By the theorem there exists an element $x \in F$ such that

$$\begin{aligned} \nu_k(x - t_k^{n_k}) &> n_k \quad \text{for } k = 1, \ldots, h \\ \nu_P(x) &\geq 0 \quad \text{for all other } P \neq Q \\ \nu_Q(x) &\geq -n. \end{aligned}$$

It follows that $\nu_k(x) = n_k$ for all $k = 1, \ldots, h$. $\square$

12.14 Residues

The final sections of this chapter are in the nature of an appendix. They discuss residues of differentials and contain a proof of the residue theorem. That makes it possible to provide a residue representation for geometric Goppa codes and show that Goppa function codes and Goppa residue codes are different representations of the same objects.

Recall that repartitions were introduced as a generalization of the concept of a function inside a closed path in C, and that differentials were introduced as a generalization of the idea of the integral of a function around such a path. We complete this analogy now by introducing the appropriate generalization of the concept of a residue and then proving the residue theorem.

DEFINITION Let P be a place and consider the repartition $1/P$ whose value at P is 1 and at any other place Q is 0. For any differential ω we define the *residue* of ω at P to be

$$\mathrm{res}_P(\omega) = \omega(1/P).$$

The following proposition assembles some useful properties of residues.

PROPOSITION *Let ω be a differential and $x \in F$.*

(a) *If $\nu_P(\omega) \geq 0$ then* $\mathrm{res}_P(\omega) = 0$.

(b) $\nu_P(\omega) \leq r$ *if and only if* $\mathrm{res}_P(\omega x) = 0$ *for all x with $\nu_P(x) \geq -r$.*

Two further statements require the additional assumption that P is a rational place.

(c) *If* $\deg(P) = 1$, $\nu_P(\omega) \geq -1$, *and* $\nu_P(x) \geq 0$, *then* $\mathrm{res}_P(\omega x) = x(P)\,\mathrm{res}_P(\omega)$.

(d) *If* $\deg P = 1$ *and* $\nu_P(\omega) = -1$, *then* $\mathrm{res}_P(\omega) \neq 0$.

Remark If $\nu_P(\omega) < -1$ or $\nu_P(\omega) = -1$ and $\deg(P) > 1$ the residue may be either zero or non-zero.

Proof (a) If $\nu_P(\omega) \geq 0$, then $1/P \in X([\omega])$ and hence $\mathrm{res}_P(\omega) = \omega(1/P) = 0$.

(b) Let $\nu_P(\omega) = s$. If $\nu_P(x) \geq -s$, then $\nu_P(\omega x) \geq 0$ and so by part (a), $\mathrm{res}_P(\omega x) = 0$. Conversely ω is not zero on $Y([\omega] + P)$ by the definition of $[\omega]$. As ω is zero on F, there exists a repartition $\Gamma \in X([\omega] + P)$ with $\omega(\Gamma) \neq 0$. Let $x = \Gamma_P$. Then $\Gamma - x1/P \in X([\omega])$ and so $\omega(\Gamma - x1/P) = 0$. Therefore

$$\mathrm{res}_P(\omega x) = \omega(x1/P) = \omega(\Gamma) - \omega(\Gamma - x1/P) = \omega(\Gamma) \neq 0.$$

(c) Since P is rational $x(P) \in K \subset F$. Furthermore, the element $x(P) - x$ of F has $\nu_P(x(P) - x) > 0$. Hence $\nu_P(\omega x - \omega x(P)) \geq 0$ and so $\mathrm{res}_P(\omega x) = \mathrm{res}_P(\omega x(P))$. Finally, $\mathrm{res}_P(\omega x(P)) = \omega(x(P)1/P)$ and as ω is K-linear, that is just $x(P)\omega(1/P)$.

(d) By part (b) there exists $x \in F$ such that $\nu_P(x) \geq 0$ and $\mathrm{res}_P(\omega x) \neq 0$. Again $\nu_P(x(P) - x) > 0$ and so $\mathrm{res}_P(\omega x(P) - \omega x) = 0$. Hence $\mathrm{res}_P(\omega x) = x(P)\,\mathrm{res}_P(\omega)$. Hence $\mathrm{res}_P(\omega) \neq 0$. □

12.15 The residue theorem

With these definitions we are now able to prove an exact analogue of the classical residue theorem which was stated in Paragraph 12.2.

THEOREM Residue theorem *Let ω be a differential and let Γ be a repartition. Then there are only finitely many places P where the residue $\mathrm{res}_P(\omega\Gamma_P)$ of the differential $\omega\Gamma_P$ is non-zero, and*

$$\omega(\Gamma) = \sum \mathrm{res}_P(\omega\Gamma_P).$$

Proof For any $x \in F$ there are only finitely many places P where $[\omega x](P) < 0$. As Γ only takes finitely many different values, it follows that there are only finitely many places where $[\omega\Gamma_P](P) < 0$. Let the set of these places be S. In order that $\mathrm{res}_P(\omega\Gamma_P) \neq 0$ we must have $1/P \notin Y([\omega\Gamma_P])$ and therefore certainly $1/P \notin X([\omega\Gamma_P])$. The only non-zero value of $1/P$ is 1 at P and so $0 + [\omega\Gamma_P](P) < 0$. Therefore S includes all those places for which $\mathrm{res}_P(\omega\Gamma_P) \neq 0$.

We now split Γ into two repartitions Δ and Λ as follows:

$$\Delta_P = \begin{cases} \Gamma_P & \text{for } P \in S \\ 0 & \text{otherwise,} \end{cases}$$

and $\Lambda = \Gamma - \Delta$.

The repartition Λ lies in $X([\omega])$ because, by construction, for $P \notin S$ we have $0 \leq [\omega\Gamma_P](P) = [\omega](P) + \nu_P(\Lambda)$ and for $P \in S$ we have $\Lambda_P = 0$. Therefore $\omega\Lambda = 0$. On the other hand,

$$\Delta = \sum_{P \in S} \Gamma_P 1/P.$$

Therefore

$$\omega(\Delta) = \sum_{P \in S} \omega(\Gamma_P 1/P) = \sum_{P \in S} \omega\Gamma_P(1/P) = \sum_{P \in S} \mathrm{res}_P(\omega\Gamma_P).$$

It follows that $\omega(\Gamma) = \omega(\Delta) = \sum \mathrm{res}_P(\omega\Gamma_P)$. The sum can be extended over all places as all other residues are 0. □

12.16 The residue representation of geometric Goppa codes

It is possible to describe the code $C_\Omega(B, D)$ directly using residues of differentials, rather than as the dual of the code $C_L(B, D)$. This description has no practical advantages, but is occasionally useful for theoretical purposes and is widely used in the literature.

PROPOSITION *Let $P_1, \ldots, P_n$ be a finite set of rational places, let $B = \sum P_i$, and let D be a divisor with support disjoint from B. Then the Goppa code $C_\Omega(B, D)$ consists of the set C of words $(\mathrm{res}_{P_1}(\omega), \ldots, \mathrm{res}_{P_n}(\omega))$ for the differentials $\omega \in J(D - B)$.*

Proof We first show that any such word is orthogonal to $C_L(B, D)$. To that end let $x \in L(D)$ and $\omega \in J(D - B)$. Then $\omega(x) = 0$ because $x \in F$, and by the residue theorem $\omega(x) = \sum \mathrm{res}_P(\omega x)$, where the sum is taken over all places. Now, if P is not in the support of B or D then $\nu_P(\omega x) \geq 0$. Similarly if P is in the support of D, then $\nu_P(x) + D(P) \geq 0$ and $\nu_P(\omega) \geq D(P)$ and so again $\nu_P(\omega x) \geq 0$. For all these places it follows that $\text{res}_P(\omega x) = 0$. For P in the support of B, $\nu_P(\omega) \geq -1$ and $\nu_P(x) \geq 0$. If either of these inequalities is strict, then $\text{res}_P(\omega x) = 0$, otherwise

$$\text{res}_P(\omega x) = \omega x(1/P) = \omega(x \cdot 1/P) = x(P)\,\text{res}_P(\omega). \qquad (*)$$

If $\nu_P(\omega) \geq 0$ then $\text{res}_P(\omega) = 0$ and if $\nu_P(x) > 0$, then $x(P) = 0$. So equation $(*)$ holds for all P in the support of B. Hence using Proposition 12.14c,

$$0 = \sum_{P \in B} \text{res}_P(\omega x) = \sum_{P \in B} x(P)\,\text{res}_P(\omega).$$

Thus we have shown that every word in C is orthogonal to all words in $C_L(B, D)$ and so C is a subset of $C_\Omega(B, D)$.

The set C is closed under addition and multiplication by constants, so we need only calculate its dimension to show that it must be the full code. Consider the map ρ taking $\omega \in J(D - B)$ to $(\text{res}_{P_1}(\omega), \ldots, \text{res}_{P_n}(\omega))$. If $\omega \in J(D)$ then $\nu_P(\omega) \geq 0$ for $P = P_1, \ldots, P_n$. Hence $\rho(\omega) = (0, \ldots, 0)$. Conversely, if $\omega \in J(D - B) \setminus J(D)$, then $\nu_P(\omega) = -1$ for some $P = P_i$ with $1 \leq i \leq n$. Hence by Proposition 12.14d, $\text{res}_P(\omega) \neq 0$ and so $\rho(\omega) \neq 0$. Thus the kernel of ρ is $J(D)$ and by the rank and nullity theorem the dimension of C is $j(D - B) - j(D)$.

In Chapter 5 we found that the dimension of $C_L(B, D)$ was $l(D) - l(D - B)$. Applying the Riemann–Roch theorem we have

$$l(D) = d(D) + j(D) + \kappa - \gamma,$$
$$l(D - B) = d(D - B) + j(D - B) + \kappa - \gamma.$$

It follows that

$$l(D) - l(D - B) = d(B) + j(D) - j(D - B) = n - \dim(C).$$

Hence C is the full orthogonal complement of $C_L(B, D)$. □

12.17 Goppa codes are dual Goppa codes

We can now prove that the class of Goppa residue codes is the same as the class of Goppa function codes. The distinction is one of presentation and not of substance.

THEOREM *Let $B = P_1 + \cdots + P_n$ be a sum of rational places. There exists a differential ω such that for $W = [\omega]$ and any divisor D with support disjoint from B, the residue code $C_\Omega(B, D)$ and the function code $C_L(B, W + B - D)$ are the same.*

Proof Let ω_0 be any non-zero differential and let $x \in F$ be such that $\nu_P(x) + \nu_P(\omega_0) = -1$ for $P \in B$ (such an x exists by the approximation theorem). Set $\omega_1 = \omega_0 x$. Then by Proposition 12.14d $\text{res}_P(\omega_1) \neq 0$ for $P \in B$. Using the approximation theorem again, there exists $y \in F$ such that $\nu_P(y - 1/\text{res}_P(\omega_1)) > 0$ for $P \in B$. It follows that $\nu_P(y) = 0$. Let $\omega = \omega_1 y$. Then $\nu_P(\omega) = -1$ and $\text{res}_P(\omega) = 1$ for $P \in B$.

Let $W = [\omega]$ and let D be a divisor with support disjoint from B. Then for $P \in B$, $W(P) + B(P) - D(P) = -1 + 1 + 0 = 0$. So $W + B - D$ and B also have disjoint support and $C_L(B, W + B - D)$ is defined. By Corollary 12.11b the map $x \mapsto \omega x$ induces an isomorphism of $L(W + B - D)$ onto $J(D - B)$ and furthermore if $x \in W + B - D$, then $\nu_P(x) \geq 0$ for all $P \in B$. Hence by Proposition 12.14c

$$\mathrm{res}_P(\omega x) = x(P)\,\mathrm{res}_P(\omega) = x(P)$$

for all $P \in B$. This shows that the code $C_L(B, W + B - D)$ consists of the words $(\mathrm{res}_{P_1}(\omega x), \ldots, \mathrm{res}_{P_n}(\omega x))$ for $x \in L(W + B - D)$. The residue representation of $C_\Omega(B, D)$ tells us that it also consists of the words of the form $(\mathrm{res}_{P_1}(\omega x), \ldots, \mathrm{res}_{P_n}(\omega x))$ for $x \in L(W + B - D)$. Thus $C_\Omega(B, D) = C_L(B, W + B - D)$. □

In the course of the proof we established that $W + B - D$ and B also have disjoint support. So we can switch the roles of D and $W + B - D$ obtaining the converse of the theorem.

COROLLARY *Under the hypotheses of the theorem we also have* $C_L(B, D) = C_\Omega(B, W + B - D)$. □

12.18 Exercises

1. Show that the sum of repartitions and the product of a repartition by a function is a repartition.

Prove that the space of all repartitions X with the defined operations forms a vector space over F and show that it is generated over K by all repartitions of the form f/S for $f \in F$ and S a set of places (both of which are allowed to vary).

2. Show that for any divisor D the sets $X(D)$ and $Y(D)$ are subspaces of the X and that every repartition lies in $X(D)$ for some positive divisor.

3. Define a product of repartitions by $(\Gamma\Delta)_P = \Gamma_P \Delta_P$ for all places. Show that this operation makes the set of repartitions into a commutative ring that is not an integral domain (and so certainly not a field).

4. Expand the proof of Proposition 12.8a. Let $\Gamma_1, \ldots, \Gamma_k$ be a basis of $X/Y(D)$. Define $\omega_i(\Gamma_j) = 1$ if $i = j$ and 0 otherwise. Show that $\omega_1, \ldots, \omega_k$ form a basis of $J(D)$.

5. Let $F = K(x)$, the field of rational functions. Show that x^{-1} is not the derivative of any element of F. Suppose further that the characteristic of K is $p \neq 0$. Show that x^{p-1} is not the derivative of any element of F. Deduce that $K(x)$ always contains differentials that do not have the form df for some $f \in F$.

6. Let $F = K(x)$ and let f be an irreducible polynomial in x. Find the divisors of df and $f dx$.

7. Let F be the function field of the curve C: $x^3 + y^3 + 1 = 0$ over F_2 and let P be the place corresponding to the point $(x{=}1, y{=}0)$. Show that $Y(P) = X$ and $Y(N) \neq X$, where N is the null divisor. Find a basis for $X/Y(N)$ and use it to find a non-zero differential ω with $[\omega] = N$.

8. Show that in the function field F of the curve C: $x^3 + y^3 + 1 = 0$ canonical divisors (that is, divisors of differentials) and principal divisors (that is, divisors of functions) are the same. Show that this can only happen in a field with $\gamma = \kappa$ (such a field is called *elliptic* if it contains a place of degree κ).
9. Let F be an elliptic function field. Show that every place is the divisor of zeros of some function.
10. Let F be an arbitrary function field and let D be a divisor with $d(D) \geq 2\gamma - 2\kappa$. Show that $j(D) = 0$ unless D is canonical, in which case $j(D) = \kappa$
11. Let the places $P_1, \ldots, P_h$ all have degree at most d and let $D = \sum_{k=1}^{h} m_k P_k$ be a positive divisor such that $d(D) > 2\gamma - 2\kappa + d$. Show that D is the divisor of zeros of some element x.
12. Let F be the function field of the Klein quartic C: $x^3y + y^3 + x = 0$ over F_2 and let P be the place corresponding to the point $(x{=}0, y{=}0)$. Show that $Y(4P) = X$ but $Y(3P) \neq X$. Deduce that $4P$ is not canonical.
13. Klein quartic (cont.). With the notation of the previous exercise show that the repartition $(1/xy)/P$ forms a basis of $X/Y(3P)$. Use this to construct a non-zero differential ω that is zero on $Y(3P)$.
14. Klein quartic (cont.). Show that if Q is the place corresponding to the point $(w{=}0, z{=}0)$ and ω is any differential that is zero on $Y(3P)$, then $\omega((y/x) \setminus Q) = 0$ and hence $\omega((y/x)/Q) = 0$. Deduce that the divisor of ω is $3P + Q$. Hence show that there is no function with divisor $P - Q$.

13

EXTENSIONS OF FUNCTION FIELDS

In this chapter we investigate how changes to a function field affect the results of Part II. The approach adopted in this book of not assuming that the base field is the field of constants simplifies the analysis somewhat. If we have a function field F'/K' extending a function field F/K we can split the extension into two parts. First we consider F'/K as an extension of F/K with the same base field and then we extend the base field K to K'.

Base field extension is straightforward and is covered in the first two paragraphs. Then I shall deal with an important special case that occurs when a curve defined over some field K is viewed over an algebraic extension field K'. In that case the function field is extended by the elements of K'. Such extensions are called constant field extensions.

The rest of the chapter (from Paragraph 13.5 onwards) deals with the general case where the base field K remains constant. The results of that part are not used in the remainder of the book and it can be omitted if you wish.

13.1 Change of base field: the genus

Our definition of the genus refers to the base field K and not to the field $\bar{K}$ of constants. It is quite possible that a field can be expressed as a function field over two different fields of constants (see Exercise 1). In that case the genera over the two fields are quite unrelated. On the other hand, if a base field K of F is contained in another base field K', then they are both contained in the same field of constants and that is the case we shall consider here. We need only investigate the extension from K to $\bar{K}$, because if we know the effect of the change from K to $\bar{K}$ and the effect of the change from K' to $\bar{K}$, we can determine the effect of the change from K to K'. All the statements of the chapters of Part II are valid for all base fields, since we made no assumptions about the base field in proving them.

In this paragraph I shall show that changing the base field from K to $\bar{K}$ merely has the effect of dividing both $d(D)$ and $l(D)$, and hence the genus γ by the constant degree $\kappa = |\bar{K} : K|$. If F is a function field over K with field of constants $\bar{K}$, then we can consider F as a function field over $\bar{K}$. We shall write $l_K(D)$, $d_K(D)$, γ_K or $l_{\bar{K}}(D)$, $d_{\bar{K}}(D)$, $\gamma_{\bar{K}}$ to distinguish the base field being referred to.

PROPOSITION (a) *For any divisor D, $l_K(D) = \kappa l_{\bar{K}}(D)$.*
(b) *For any divisor D, $d_K(D) = \kappa d_{\bar{K}}(D)$.*
(c) $\gamma_K = \kappa\gamma_{\bar{K}}$.

Proof (a) If $\alpha_1, \ldots, \alpha_r$ form a basis of $L(D)$ over $\bar{K}$ and $v_1, \ldots, v_\kappa$ form a basis of $\bar{K}$ over K, then the products $a_i v_j$ form a basis of $L(D)$ over K.

(b) We need only prove the degree statement for a single place P. By definition $d_K(P) = |F_P/P : K|$. By Corollary 9.7b F_P/P is a field containing $\bar{K}$.

$$d_K(P) = |F_P/P : K| = |F_P/P : \bar{K}|\,|\bar{K} : K| = d_{\bar{K}}(P)\kappa.$$

(c) By the definition of the genus $l_{\bar{K}}(D) \geq d_{\bar{K}}(D) + |\bar{K} : \bar{K}| - \gamma_{\bar{K}}$ for all divisors with equality for some divisor D, where of course $|\bar{K} : \bar{K}| = 1$. We rewrite this as $l_{\bar{K}}(D) - d_{\bar{K}}(D) \geq 1 - \gamma_{\bar{K}}$. From parts (a) and (b) it follows that

$$l_K(D) - d_K(D) = \kappa(l_{\bar{K}}(D) - d_{\bar{K}}(D)) \geq \kappa(1 - \gamma_{\bar{K}}),$$

for all divisors D with equality for some D. On the other hand we also have

$$l_K(D) - d_K(D) \geq \kappa - \gamma_K,$$

for all divisors D with equality for some D. Therefore $\kappa - \gamma_K = \kappa(1 - \gamma_K)$ and it follows that $\gamma_K = \kappa\gamma_{\bar{K}}$. □

COROLLARY *Genus 0 depends only on $\bar{K}$. Genus 1 can only occur if $K = \bar{K}$.* □

13.2 Change of base field: differentials

We now consider the effect of changing the base field from K to $\bar{K}$ on repartitions and differentials. The definitions of repartitions and the spaces $X(D)$ and $Y(D)$ are independent of the base field so that their dimensions are automatically divided by the constant degree κ. On the other hand, the definition of a differential explicitly invokes the base field and it is never the case that a non-zero K-differential is a $\bar{K}$-differential, unless $K = \bar{K}$. Nevertheless, as you shall see, both K-differentials and $\bar{K}$-differentials define the same canonical divisors, so these are also independent of the base field.

In order to study the relation between K-differentials and $\bar{K}$-differentials we choose an arbitrary non-zero K-linear map π from $\bar{K}$ onto K. We shall assume that π restricts to the identity on K itself. Numerous such maps can be constructed, for example by choosing a K-basis $1 = \alpha_1, \ldots, \alpha_n$ of $\bar{K}$ and mapping $\sum a_i\alpha_i$ to a_1.

For any $\bar{K}$-differential ω, the map $\pi\omega$ from X to K which maps the repartition Γ to $\pi(\omega(\Gamma))$ is a differential, because it is K-linear and zero on $Y(D)$ if ω is zero on $Y(D)$. In the following proposition I shall show that the divisor of $\pi\omega$ is the same as the divisor of ω, proving that canonical divisors are independent of the base field.

PROPOSITION *Let $\pi\omega$ be defined as above. If $\omega \neq 0$, then $\pi\omega \neq 0$. Furthermore $\pi\omega$ is zero on $Y([\omega])$ but not zero on $Y([\omega] + P)$ for any place P.*

Proof Certainly the fact that ω is zero on $Y([\omega])$ ensures that $\pi\omega$ is also zero on $Y([\omega])$. Let P be any place. Then from the fact that ω is zero on F and not zero on $Y([\omega]+P)$ it follows that there exists $\Gamma \in X([\omega]+P)$ such that $\omega(\Gamma) \neq 0$. Then $\Delta = (1/\omega(\Gamma))\Gamma$ also lies in $X([\omega]+P)$ and $\omega(\Delta) = 1$ which implies $\pi\omega(\Delta) = 1$. Thus $\pi\omega$ is not zero on $Y([\omega]+P)$, and incidentally $\pi\omega \neq 0$. □

COROLLARIES (a) *The divisors* $[\omega]$ *and* $[\pi\omega]$ *are the same.*
(b) *In changing from* K *to* $\bar{K}$ *all the terms in the Riemann–Roch theorem are divided by* κ*; the divisors remain unchanged.* □
This gives a second proof that $\gamma_K = \kappa\gamma_{\bar{K}}$.

13.3 Constant field extensions

In the applications of function fields it frequently occurs that one wishes to extend a function field F/K so that the base field K satisfies certain additional conditions. That motivates the following definition.

DEFINITION Let F be a field of functions over K and let $\alpha_1, \ldots, \alpha_n$ be elements of an extension field of K that are algebraic over K; then putting $F' = F(\alpha_1, \ldots, \alpha_n)$ and $K' = K(\alpha_1, \ldots, \alpha_n)$, the function field F' over K' is called a *constant field extension* of the function field F over K.

It is easy to verify that F' really is a field of functions over K and that K' is contained in its field of constants.

To simplify the arguments we make an additional assumption that always holds for the fields used in coding theory.

ASSUMPTION The field F contains a rational place (that is, a place of degree 1).

This assumption has the consequence that K must be the field of constants of F. It is possible to determine what happens to extensions without this assumption and it turns out that for finite or perfect fields the genus over the full field of constants always remains fixed under a constant field extension. For non-perfect fields it may change. For more detail on the arguments, which are technically quite intricate, the reader is referred to the books of Stichtenoth (1991) which covers the perfect case and Deuring (1973) which deals with the general case.

In the proof that the genus remains constant we shall make use of the following lemma and its corollaries.

LEMMA *Let* F *be a field of functions over the field* K *with a rational place* P. *Let* K' *be a finite extension of* K *and let* $\alpha_1, \ldots, \alpha_n$ *be a vector space basis of* K' *over* K. *Let* $F' = F(\alpha_1, \ldots, \alpha_n)$ *be the corresponding constant field extension and let* P' *be a place of* F' *such that* $P' \cap F = P$. *Then for any functions* $x_1, \ldots, x_n \in F$,

$$\nu_{P'}\left(\sum_{i=1}^{n} \alpha_i x_i\right) = \min\{\nu_P(x_i)\}.$$

Proof Since the elements α_i are algebraic over K they are constants and therefore $\nu_{P'}(\alpha_i) = 0$. By the general properties of places, we have

$$\nu_{P'}\left(\sum_{i=1}^{n} \alpha_i x_i\right) \geq \min\{\nu_{P'}(x_i)\}. \qquad (*)$$

We must show that equality holds in the inequity $(*)$ and that $\nu_{P'}$ can be replaced by ν_P on the right hand side. Suppose therefore that $\nu_{P'}(x_1)$ is minimal and that the inequality in eqn $(*)$ is strict. Then

$$\nu_{P'}\left(\sum \alpha_i \frac{x_i}{x_1}\right) > 0.$$

By the assumption x_i/x_1 lies in the place ring $F_P \subseteq F'_{P'}$ of P. Taking residues modulo the maximal ideal P' we obtain

$$\sum \alpha_i \overline{\left(\frac{x_i}{x_1}\right)} = 0. \qquad (**)$$

Since P is rational, $\overline{(x_i/x_1)} \in K$ and in particular $\overline{x_1/x_1} = 1$. But then eqn $(**)$ contradicts the basis property of $\alpha_1, \ldots, \alpha_n$. So the assumption of strict inequality in $(*)$ is untenable.

To show that we can replace $\nu_{P'}$ by ν_P on the right hand side of $(*)$ we shall show that $\nu_{P'}(y) = \nu_P(y)$ for all $y \in F$. Choose an element $x \in F$ with $\nu_P(x) = 1$. Then $\nu_{P'}(x) = k \geq 1$. Every element $y \in F$ such that $\nu_P(y) = n$ can be written in the form ux^n, where $\nu_P(u) = 0$. By definition neither u nor $1/u$ has a zero at P and so it follows that $\nu_{P'}(u) = 0$. Hence

$$\nu_{P'}(y) = 0 + n\nu_P(x) = k\nu_P(y).$$

Now choose an element $y' = \sum \alpha_i y_i \in F'$ such that $\nu_{P'}(y') = 1$. By eqn $(*)$,

$$1 = \nu_{P'}(y') = \min\{\nu_{P'}(y_i)\} = k \cdot \min\{\nu_P(y_i)\}.$$

That is only possible if $k = 1$. Then $\nu_{P'}(y) = \nu_P(y)$ for all $y \in F$ and the lemma is established. □

COROLLARIES (a) *If P is a rational place of F then there is a unique place P' of F' extending P and the field of constants of F' is K'.*
(b) *Under the assumptions of the lemma, the elements $\alpha_1, \ldots, \alpha_n$ form a vector space basis of F' over F.*

Proofs (a) We have found that the valuation corresponding to P' is fixed. Since different places have different valuations, there can be only one place. The proof of the lemma shows that the residue field of P' is K'. Every residue field of a place contains the field of constants of F' so K' contains the field of constants. But K' is algebraic over K and so K' is contained in the field of constants.

(b) It follows directly from the definition that every element of F' can be written in the form $x_1\alpha_1 + \cdots + x_n\alpha_n$ with $x_i \in F$. It remains to show that $\alpha_1, \ldots, \alpha_n$ are linearly independent. We must show that $x_1\alpha_1 + \cdots + x_n\alpha_n = 0$ with $x_i \in F$ implies that $x_i = 0$ for all $i = 1, \ldots, n$.

If not all of $x_1, \ldots, x_n$ are zero then $\min\{\nu_P(x_i)\}$ is finite and so by the lemma $\nu_{P'}(\sum \alpha_i x_i) < \infty$. Then the sum cannot be zero. □

13.4 The genus of constant field extensions

Before using the lemma above to determine the genus of a constant field extension we look at an example.

EXAMPLE Consider the place corresponding to the point P: $(0,1)$ of the function field of the cubic curve C: $x^3 + y^3 + 1 = 0$ over finite extensions of $K = \mathsf{F}_2$. The relationship between points and places in general is discussed in the next chapter, but we found in Example 3.13 that the x-order for this point is a discrete valuation and you can check that the local ring $K[C]_P$ is a place.

In Example 4.6 we found that $\nu_P(x^i y^j/(y+1)^{i+j}) = -2i - 3j$ and that the functions

$$1,\ x/(y+1),\ y/(y+1),\ x^2/(y+1)^2, xy/(y+1)^2, x^3/(y+1)^3, \ldots$$

form successive bases of $L(mP)$ over K. The argument used to establish this fact was independent of the base field used and so would still apply over a finite extension $K' = \mathsf{F}_{2^n}$ of K.

If. however, we are considering the field $K'(C)$ as a field of functions over K, then we have to rewrite linear combinations $\sum \beta_{ij} x^i y^j/(y+1)^{i+j}$ that have coefficients in K' as linear combinations with coefficients in K. For that purpose we introduce a basis $\alpha_1, \ldots, \alpha_n$ of K' over K and write $\beta_{ij} = b_{ij1}\alpha_1 + \cdots + b_{ijn}\alpha_n$. In that way we can rewrite our linear combination as a combination of the elements $\alpha_k x^i y^j/(y+1)^{i+j}$ and so the dimension of $L(mP)$ is multiplied by n.

In the general case we do not have the benefit of calculations that are known to be independent of the base field. So we follow the thread of this example in the reverse direction.

PROPOSITION *Let F/K be a field of functions with a rational place P and let F' be a constant field extension. If the field of constants of F' is K', then the genera γ and γ' of F and F' over K satisfy $\gamma' = |K' : K|\gamma$.*

Proof Let P' be the place of F' lying over P. We shall show that $\kappa l_K(mP) = l_K(mP')$ for all m.

Let $\alpha_1, \ldots, \alpha_\kappa$ be a basis of K' over K. Observe that since $\nu_{P'}(\sum \alpha_i x_i) = \min\{\nu_P(x_i)\}$, it follows that $\sum \alpha_i x_i \in L(mP')$ if and only if $x_i \in L(mP)$ for all i. Let $y_1, \ldots, y_l$ be a K-basis of $L(mP)$. Every element of $L(mP)$ can be written as a linear combination of $y_1, \ldots, y_l$ with coefficients in K and every element of $L(mP')$ can be written as a sum $\sum \alpha_i x_i$ with $x_i \in L(mP)$. Gathering terms, it follows that every element of $L(mP)$ can be written as a linear combination of the terms $\alpha_i y_j$ with coefficients in K.

We must also show that the elements $\alpha_i y_j$ are linearly independent over K. If that is not the case, then

$$\sum_{i=1}^{\kappa}\sum_{j=1}^{l} b_{ij}\alpha_i y_j = 0,$$

for some $b_{ij} \in K$ not all zero. Changing the order of summation we obtain

$$\sum_i \alpha_i\Big(\sum_j b_{ij}y_j\Big) = 0.$$

Therefore by Corollary 13.3b $\sum b_{ij}y_j = 0$ for all i. The basis property of $y_1, \ldots, y_l$ over K then implies that $b_{ij} = 0$ for all i, j contradicting our assumption. That establishes that $\dim L(mP') = \kappa \dim L(mP)$.

Using the fact that K is the field of constants of F and K' is the field of constants of F', the Riemann–Roch theorem gives

$$\gamma' = l(mP') - d(mP') - \kappa$$
$$\gamma = l(mP) - d(mP) - 1$$

for large m. By assumption P is rational and so $d(mP) = m$ and by Corollary 13.3a the residue field of P' is K' and so $d(P') = \kappa$. Hence $d(mP') = m\kappa$.

We have proved above that $l(mP') = \kappa l(mP)$ and so $\gamma' = \kappa\gamma$ as claimed. □

COROLLARY *The genus of F' over K' is γ.*

Proof This follows directly from the proposition and Proposition 13.2. □

13.5 Places in an extension field

We now turn to the general case where F is a function field over K and F' is a finite algebraic extension of F. Every function field is obtained from a rational function field in this manner, so the investigation of such extensions is quite natural. It is straightforward to verify that F' is still a function field over K and the existence theorem guarantees that every place of F extends to a place of F'. Conversely $P' \cap F$ is a place of F for any place P' of F'. So we begin by investigating the places lying above a particular place of F.

If P' lies above P, then the residue field of $F'_{P'}$ contains the residue field of F_P as a subfield, and $\nu_P(x) > 0$ implies $\nu_{P'}(x) > 0$ for $x \in F$, though it may not be true that $\nu_{P'}(x) = \nu_P(x)$.

DEFINITION Let P' be a place of F' and let $P = P' \cap F$. We define the *relative degree* $f(P')$ of P' to be the degree of the residue field of $F'_{P'}$ over that of F_P:

$$f(P') = |F'_{P'}/P' : F_P/P|.$$

We define the *ramification index* $e(P')$ of P' to be the minimum value $\nu_{P'}(x)$ over all x in P:

$$e(P') = \min\{\nu_{P'}(x) : \nu_P(x) > 0\}.$$

EXAMPLE Consider as F the field of rational functions $\mathsf{R}(x)$ over the real numbers and as F' the field of rational functions $\mathsf{C}(y)$ with $y^2 = x$, taken as a function field over R.

The finite places of F correspond to irreducible real polynomials in x and those of F' correspond to linear complex polynomials in y, because C is obviously the field of constants of F'.

Let P be the place corresponding to x. There is only one place P' of F' lying above P, namely the place corresponding to y. Its residue field is C which has degree 2 over R so $f(P') = 2$. The element y is a local parameter for P' and so $e(P') = 2$ also.

Let Q be the place corresponding to $x - 1$. There are two places Q' lying above Q. They correspond to $y \pm 1$. For both of them the residue field is again C and so $f(Q') = 2$. If Q' corresponds to, say, $y - 1$, then the equation $x - 1 = (y + 1)(y - 1)$ shows that $\nu_{Q'}(x - 1) = 1$ and so $e(Q') = 1$.

Let R be the place in F corresponding to $x^2 + 1$. There are four places R' lying above R corresponding to the polynomials $y \pm \sqrt{\pm\mathsf{i}}$. As the residue field of R is C each of these places has $f(R') = 1$ and the formula

$$x^2 + 1 = (y + \sqrt{\mathsf{i}})(y - \sqrt{\mathsf{i}})(y + \sqrt{-\mathsf{i}})(y - \sqrt{-\mathsf{i}})$$

shows that over each $\nu_{R'}(x^2 + 1) = 1$ and so $e(R') = 1$.

Before investigating the patterns suggested by this example we prove some elementary properties of the relative degree and ramification index.

PROPOSITION *Let F' be a finite extension of a function field F over K and let P' be a place of F'. Denote by e and f the ramification index and relative degree of P'. If $P = P' \cap F$ then $d(P') = fd(P)$ and $\nu_{P'}(x) = e\nu_P(x)$ for all $x \in F$.*

Proof The first statement follows directly from the product formula for the degrees of field extensions. By definition $d(P) = |F_P/P : K|$ and so

$$fd(P) = |F'_{P'}/P' : F_P/P|\,|F_P/P : K| = |F'_{P'}/P' : K| = d(P').$$

To prove the second statement, let t be a local parameter of P and define e' to be $\nu_{P'}(t)$. Then any $x \in F$ can be written in the form at^n, where $\nu_P(a) = 0$ and $n = \nu_P(x)$. It follows that $\nu_{P'}(a) = 0$ and so $\nu_{P'}(x) = ne'$. Therefore the minimum value of $\nu_{P'}(x)$ is e' and thus $e = e'$. That establishes the claim. □

13.6 Extensions of a place

The patterns of Example 13.5 might lead one to make three conjectures (beware, patterns can be misleading):

1. there should only be finitely many places above a given place;
2. they should all have the same relative degree and ramification index; and
3. the sum of the values $e_i f_i$ over these places should be the degree $|F' : F|$ of the extension field.

The first and third of these statements are true, but the second is not.

EXAMPLE Let F be the field of rational functions $\mathsf{Q}(x)$ over the rational numbers Q and let $F' = \mathsf{Q}(\alpha, y)$ where $\alpha^4 = 3$ and $y^2 = x$. Then the field of constants of F' is $\mathsf{Q}(\alpha)$ which has degree 4 over Q.

Consider the place P corresponding to $x^2 - 3$ in F. There are three places P'_k lying above P in F', which correspond to the polynomials $y + \alpha$, $y - \alpha$, and $y^2 + \alpha^2$. As $x^2 - 3$ is the product of these three polynomials all three places have ramification index 1, but the relative degrees are different, because the residue fields of the first two places are $\mathsf{Q}(\alpha)$ while the residue field of the third place is $\mathsf{Q}(\alpha, \mathrm{i})$. Thus the relative degrees are 2, 2, and 4.

The construction of an example with varying ramification indices is left to Exercise 4.

PROPOSITION *Let F' be a finite algebraic extension of the function field F over K and let P be a place of F; then there are only finitely many places P' of F' such that $P' \cap F = P$.*

If these places are $P'_1, \ldots, P'_k$, then

$$\sum_{i=1}^{k} e(P'_i) f(P'_i) = |F' : F|.$$

Proof For sufficiently large m the space $L(mP)$ has dimension $l(mP) > 0$ and so there exists an element $x \in F$ with P as its only pole. Let the divisor of poles of x be $[x]_\infty = rP$. The degree theorem tells us that $d([x]_\infty) = |F : K(x)|$.

Then the poles of x in F' are the places extending P. By the zeros theorem there are only finitely many such places. So we can number them $P'_1, \ldots, P'_k$. The degree of P'_i is $d(P'_i) = f(P_i) d(P)$ and the valuation of x at P'_i is $e(P'_i)\nu_P(x) = e(P'_i) r$. So the divisor of poles of x in F' is $r \sum e(P'_i) P'_i$ which has degree $r \sum e(P'_i) f(P'_i)$. Applying the degree theorem to x as an element of F' we find

$$|F' : K(x)| = r \sum_{i=1}^{k} e(P'_i) f(P'_i) = |F : K(x)| \sum_{i=1}^{k} e(P'_i) f(P'_i).$$

As $|F' : K(x)| = |F' : F|\,|F : K(x)|$, the claim follows. □

13.7 Extending divisors

In the light of this theorem it makes sense to define the extension of the P to F' to be the sum $\sum e(P'_i) P'_i$.

DEFINITION Let F' be a finite extension of the function field F. For any place P of F we define the *extension* of P to F' to be the divisor

$$P^{F'} = \sum_{P' \cap F = P} e(P')P'.$$

For any divisor $D = \sum D(P)P$ of F we define the *extension* of D to F' as $D^{F'} = \sum D(P)P^{F'}$.

PROPOSITION *Let F' be a finite extension of the function field F and let D be a divisor of F. Then $D^{F'}$ is a divisor of F'. Furthermore, $d(D)^{F'} = |F' : F|\, d(D)$ and $L(D^{F'}) \cap F = L(D)$.*

Proof There are only finitely many places of F where $D(P) \neq 0$ and there are only finitely many places of F' above each of these. So $D^{F'}(P') \neq 0$ for only finitely many places of F' and thus $D^{F'}$ is a divisor.

Proposition 13.6 can be restated in the form $d(P^{F'}) = |F' : F|\, d(P)$ and the statement about the degree of $D^{F'}$ then follows immediately.

For $x \in F$ the condition $x \in L(D)$ is equivalent to $\nu_P(x) + D(P) \geq 0$ for all P. For any place P' of F' with $P' \cap F = P$ that is equivalent to $e(P')\nu_P(x) + e(P')D(P) \geq 0$. From the definition $e(P')D(P) = D^{F'}(P')$ and we have proved that $\nu_{P'}(x) = e(P')\nu_P(x)$. So the condition is equivalent to $\nu_{P'}(x) + D^{F'}(P) \geq 0$. Extended over all places of F' that is precisely the condition that $x \in L(D^{F'})$. □

13.8 Affine rings of extension fields

The method we shall use to calculate the genus of an extension field relies on some subtle analysis of the way the local ring of a point extends to F'. Part of this analysis relies on an investigation of the way another ring, the affine ring, of a place extends to F'. I have indented some of the more technical proofs of this section to indicate that you can read on without digesting their details.

DEFINITION Let P be a place of the function field F. The *affine* ring of P denoted by F^P consists of the elements of F that have P as their only pole.

There are two alternative characterizations of F^P, namely as the intersection of the local rings F_Q over all places Q different from P and also the union of the spaces $L(nP)$ for all $n \geq 0$. The first of these characterizations has as an immediate consequence that F^P is a ring. The second characterization shows that F^P contains non-constant elements, because $l(nP) > \kappa$ for sufficiently large values of n.

The *extension* of F^P to F' is defined in the natural way.

DEFINITION Let P be a place of a function field F and let F' be a finite extension of F. The extension F'^P of F^P to F' consists of all elements of F' with poles only at places P' such that $P' \cap F = P$.

PROPOSITION *Let P be a place of the function field F and let F' be a finite extension of F. Then there exist finitely many elements $\eta_1, \ldots, \eta_r$ of F'^P such that every element of F'^P can be represented as $\sum y_i\eta_i$ with $y_i \in F^P$ for all i.*

In other words F'^P is a finitely generated F^P-module.

Proof Let x be a non-constant element of F^P and let D' be the divisor of poles of x in F' (so $D' = [x]_\infty^{F'}$). Thus D' is positive and $D'(P') > 0$ for precisely those places P' of F' with $P' \cap F = P$. Any element of F'^P is contained in $L(nD')$ for some n and certainly $L(nD') \subseteq F'^P$. So F'^P is the union of the spaces $L(nD')$ for positive n.

It follows from Riemann's theorem that for sufficiently large n

$$l((n+1)D') - l(nD') = d(D').$$

Fix such an n. If $\eta_1, \ldots, \eta_d$ is a basis of $L((n+1)D')$ modulo $L(nD')$, then $x^k\eta_1, \ldots, x^k\eta_d$ is a basis of $L((n+k+1)D')$ modulo $L((n+k)D')$. Hence for every element $\eta \in F'^P$ there exist polynomials $f_i(x) \in K[x]$, where K is the base field of F, such that

$$\eta - \sum_{i=1}^{d} f_i(x)\eta_i \in L(nD').$$

If we let $\eta_{d+1}, \ldots, \eta_r$ be a basis of $L(nD)$, then it follows that

$$\eta = \sum_{i=1}^{d} f_i(x)\eta_i + \sum_{i=d+1}^{r} a_i\eta_i$$

with $f_i(x) \in K[x]$ and $a_i \in K$. As $K \subset K[x] \subseteq F^P$ this proves the claim.

□

13.9 Coordinate tests

Just as we can extend the affine ring F^P to F' we can also extend the place ring F_P.

DEFINITION Let P be a place of the function field F and let F' be a finite extension of F. Then the *extension* F'_P of F_P to F' is the intersection of the place rings $F'_{P'}$ for all places P' of F' with $P' \cap F = P$.

The proposition we have just proved allows us to test for membership of F'^P and F'_P in terms of the coordinate representation of an element with respect to an arbitrary basis of F' over F.

PROPOSITION (Chevalley) *Let P be a place of a function field F and let the finite extension F' of F have the basis $\xi_1, \ldots, \xi_n$ over F.*

(a) *There exists an element $v \in F$ such that $\sum_{i=1}^n y_i\xi_i \in F'^P$ implies that $vy_i \in F^P$ for all i.*

(b) *For any place $Q \neq P$ of F furthermore, $\sum_{i=1}^n y_i\xi_i \in F'_Q$ implies that $vy_i \in F_Q$ for all i (with the same element v).*

Proof (a) Let $\eta_1, \ldots, \eta_r$ have the property that $\eta \in F'^P$ has a representation $\eta = \sum_{j=1}^r f_j\eta_j$ with $f_j \in F^P$ for all j. Represent $\eta_j = \sum_{i=1}^n e_{ij}\xi_i$ in terms of the basis $\xi_1, \ldots, \xi_n$. Let $Q_1, \ldots, Q_k$ be the places of F different from P that are poles of at least one e_{ij}. Then by the approximation theorem there exists $v \in F^P$ such that $\nu_{Q_l}(v) \geq -\nu_{Q_l}(e_{ij})$ for all i, j, and l. The element $\eta \in F'^P$ satisfies

$$\eta = \sum_{j=1}^r f_j\eta_j = \sum_{i=1}^n \Big(\sum_{j=1}^r f_j e_{ij}\Big)\xi_i$$

with $f_j \in F^P$. As $\xi_1, \ldots, \xi_n$ is assumed to be a basis $\eta = \sum y_i\xi_i$ implies that $y_i = \sum_j f_j e_{ij}$. By construction $ve_{ij} \in F^P$ and $f_j \in F^P$; hence $vy_i = v\sum_{j=1}^r f_j e_{ij} \in F^P$ for all $i = 1, \ldots, n$.

(b) Let $\eta = \sum_{i=1}^n y_i\xi_i$ lie in F'_Q and let $Q'_1, \ldots, Q'_r$ be the poles of η that do not intersect F in P and let $Q_i = Q'_i \cap F$ (it is possible that a place is repeated in this list). No pole Q' of η has $Q' \cap F = Q$ and so using the approximation theorem again we can find $z \in F^P$ such that $\nu_Q(z) = 0$ and $\nu_{Q_l}(z) \geq -\nu_{Q'_l}(\eta)$ for $l = 1, \ldots, r$. Then $z\eta \in F'^P$ and so by part (a) $vzy_i \in F^P \subset F_Q$ for all $i = 1, \ldots, n$. Since $\nu_Q(z) = 0$, it follows that $z^{-1} \in F_Q$ and so $vy_i \in F_Q$ for all i. □

The proposition has the following useful corollary.

Corollary *Let $\xi_1, \ldots, \xi_n$ be a basis of the extension field F' over the function field F. Then at all places Q with a finite number of exceptions, $\eta = \sum_{i=1}^n y_i\xi_i$ lies in F'_Q if and only if y_i lies in F_Q for $i = 1, \ldots, n$.*

Proof If ξ_i lies in F'_Q for all i, then the condition is sufficient. There are only finitely many places Q' for which $\xi_i \notin F'_{Q'}$ and so the condition is sufficient for all but the places $Q' \cap F$ for these Q'.

With v as in the proposition $vy_i \in F'_Q$ implies $y_i \in F'_Q$ provided that $v^{-1} \in F'_Q$. There are only finitely many places Q' where $v^{-1} \notin F'_{Q'}$ and so the condition is necessary for all places of F except P and the places $Q' \cap F$ for these Q'. □

The basis property described by the corollary above will turn out to be of some significance. So we introduce a name for such bases.

Definition The basis $\xi_1, \ldots, \xi_n$ of F' over F is called *integral* over Q if $\eta = \sum_{i=1}^n y_i\xi_i$ lies in F'_Q if and only if y_i lies in F_Q for $i = 1, \ldots, n$.

13.10 Integral bases

We have seen that every basis of F' over F is integral for all but a finite number of places P of F. We shall now show that integral bases can be constructed for all places.

Proposition *Let P be a place of the function field F and let F' be a finite extension of F. Then there exists a basis $\xi_1, \ldots, \xi_n$ of F' over F that is integral over P.*

Proof Let $\eta_1, \ldots, \eta_n$ be any basis of F' over F and consider the elements $\eta \in F'_P$. Writing $\eta = \sum y_i\eta_i$ we find that the value $\nu_P(y_i)$ is bounded below because there exists a fixed element v such that $\nu_P(vy_i) \geq 0$ for all such η. So we choose $\xi_1 = \sum x_i\eta_i$ in F'_P so that the value $\nu_P(x_1)$ is as small as possible. By Exercise 6, $x_1 \neq 0$, so we consider the new basis $\xi_1, \eta_2, \ldots, \eta_n$.

For any $\eta \in F'_P$ with

$$\eta = u_1\xi_1 + \sum_{i=2}^{n} u_i\eta_i$$

we have

$$\eta = u_1x_1\eta_1 + \sum_{i=2}^{n}(u_i + u_1x_i)\eta_i.$$

By the construction of x_1 we have $\nu_P(u_1x_1) \geq \nu_P(x_1)$ and therefore $\nu_P(u_1) \geq 0$. Furthermore

$$\nu_P(u_i) = \nu_P(u_i + u_1x_i - u_1x_i) \geq \min(\nu_P(u_i + u_1x_i), \nu_P(x_i)).$$

So the minimum P-valuation for each of the coefficients of η in terms of the new basis has not decreased in comparison with the old basis.

We can repeat this process for each element η_i and end up with a basis $\xi_1, \ldots, \xi_n$ consisting of elements of F'_P such that every $\eta \in F'_P$ is represented as $\eta = \sum_{i=1}^{n} u_i\xi_i$ with $\nu_P(u_i) \geq 0$ for all i. That is the required integral basis over P. □

13.11 Repartitions of an extension

We shall estimate the genus of F' from the genus of F by constructing a differential of F' from one of F. The degrees of their canonical divisors will then determine the genera. To that end we must investigate the sets of repartitions of F' and of F, which we denote by X' and X. When we extend from F to F' it is difficult to distinguish between places that lie above the same place of F so (following an idea of P. Roquette) we introduce a special class of repartitions of F' which we denote by X''.

DEFINITION A repartition Γ' of F' is said to be *relatively constant* if $\Gamma'_{P'} = \Gamma'_{P''}$ whenever $P' \cap F = P'' \cap F$. We denote the set of relatively constant repartitions by X''.

The relatively constant repartitions form a proper subspace of the set of all repartitions of F', but it is quite close to the whole set, as the following proposition shows.

PROPOSITION *Let F' be a finite extension of the function field F. Let the repartitions of F' be denoted by X' and the relatively constant repartitions by X''. Then $X' = X'' + X'(D')$ for any divisor D' of F'.*

Proof Recall that $X'(D')$ consists of all repartitions Γ' such that $\nu_{P'}(\Gamma') + D'(P') \geq 0$ at all places P' of F'.

Let $\xi \in F'$ and let S' be an arbitrary set of places of F'. We need only show that $\Gamma' = \xi/S'$ lies in $X'' + X'(D')$. Let T_0' be the set of places Q' for which

$$\nu_{Q'}(\Gamma') + D'(Q') < 0.$$

For such places we must have $\nu_{Q'}(\xi) < 0$ or $D'(Q') < 0$. So T_0' is finite. Now let T' be the set of places P' such that $P' \cap F = Q' \cap F$ for some $Q' \in T_0'$. There are only finitely many P' for each Q'. So T' is finite also.

By the approximation theorem there exists $\eta \in F'$ such that

$$\nu_{P'}(\Gamma'_{P'} - \eta) + D'(P') \geq 0$$

for all $P' \in T'$. Let Δ' be the repartition η/T'. Then $\Delta' \in X''$ and for $P' \in T'$ we have $\nu_{P'}(\Gamma'_{P'} - \Delta'_{P'}) + D'(P') \geq 0$. For Q' outside T' we have $\Delta'_{Q'} = 0$ and $\nu_{Q'}(\Gamma') + D'(Q') \geq 0$, so there also $\nu_{Q'}(\Gamma'_{Q'} - \Delta'_{Q'}) + D'(Q') \geq 0$. Hence $\Gamma' - \Delta' \in X'(D')$. Thus $\Gamma' = \Delta' + (\Gamma' - \Delta')$ lies in $X'' + X'(D')$ completing the proof. □

13.12 The trace

Our construction of a differential in F' requires the use of a non-zero F-linear map from F' to F. Traditionally the map used for this purpose has been the trace. This has the advantage that a lot is known about this map, but the disadvantage that the trace is sometimes zero.

In order to define the trace of an element ξ of F' we consider the map from F' to itself defined by taking every element η to $\xi\eta$. This map is certainly F-linear, but of course it does not take F' to F. Choose any basis $\xi_1, \ldots, \xi_n$; then multiplication by ξ is represented by a matrix M_ξ, where the (i, j) entry is the coefficient of ξ_i in the sum $\xi\xi_j = \sum m_{ij}\xi_i$. The matrix $M_{\xi+\zeta}$ of multiplication by $\xi + \zeta$ is the sum of the matrices M_ξ and M_ζ, and the matrix $M_{u\xi}$ of multiplication by $u\xi$ is uM_ξ for any u in F.

Definition We define the *trace* $\mathrm{Tr}(\xi)$ of ξ to be the trace of M_ξ, that is the sum $\sum_{i=1}^{n} m_{ii}$ of its diagonal elements. If it is necessary to specify the fields we write $\mathrm{Tr}_{F'/F}(\xi)$. The extension F' of F is called *separable* if $\mathrm{Tr}(\zeta)$ is non-zero for some element of $\zeta \in F'$.

Example Consider the field C over R and the two bases $1, \mathsf{i}$ and $1 + \mathsf{i}, 1 - 2\mathsf{i}$. Let $\xi = 1 + 2\mathsf{i}$; then the matrices of multiplication by ξ with respect to these two bases are

$$\begin{bmatrix} 1 & -2 \\ 2 & 1 \end{bmatrix} \text{ and } \begin{bmatrix} 1/3 & -4/3 \\ 10/3 & 5/3 \end{bmatrix}$$

Notice that both matrices have trace 2.

The following proposition states the most important properties of the trace of field elements.

PROPOSITION *Let F' be a finite extension of the field F.*
(a) *The trace of ξ does not depend on the choice of basis used to calculate the multiplication matrix of ξ.*
(b) *If F'' is a finite extension of F' and $\xi \in F''$, then*

$$\mathrm{Tr}_{F''/F}(\xi) = \mathrm{Tr}_{F'/F}(Tr_{F''/F'}(\xi)).$$

(c) *If F is a function field and P is a place of F then for $\xi \in F'_P$ the trace* $\mathrm{Tr}(\xi)$ *lies in F_P.*

Proof (a) This follows from the standard properties of the trace of a matrix proved in linear algebra.

Let $\xi_1, \ldots, \xi_n$ and $\eta_1, \ldots, \eta_n$ be two bases of F' over F and let H be the matrix with entries h_{ij} where $\eta_j = \sum_i h_{ij}\xi_i$. Then H is invertible and its inverse is defined in the same way by exchanging the roles of the two bases. Direct matrix calculations show that for any element ζ whose multiplication matrix with respect to $\xi_1, \ldots, \xi_n$ is M the multiplication matrix with respect to $\eta_1, \ldots, \eta_n$ is $H^{-1}MH$.

In linear algebra the *characteristic polynomial* of an $n \times n$ matrix M is defined as $\det(\dot{x}I - M)$. Again direct calculation shows that the trace of M is -1 times the coefficient of $\dot{x}^{n-1}$ in the characteristic polynomial of M. As $\dot{x}I - H^{-1}MH = H^{-1}(\dot{x}I - M)H$ and the determinant of H^{-1} is the inverse of the determinant of H, it follows that M and $H^{-1}MH$ have the same characteristic polynomial and hence also the same trace.

(b) Let $\xi_1, \ldots, \xi_n$ be a basis of F' over F and $\eta_1, \ldots, \eta_m$ be a basis of F'' over F'. Then the set $\{\xi_r\eta_i\}$ forms a basis of F'' over F. Let $\zeta \in F''$ have as its multiplication matrix with respect to $\eta_1, \ldots, \eta_m$ the matrix (μ_{ij}) and let μ_{ij} have as its multiplication matrix with respect to $\xi_1, \ldots, \xi_n$ the matrix (x^{ij}_{rs}), where i, j are fixed and r, s vary from 1 to n. Then the multiplication matrix of ζ with respect to the basis $\xi_r\eta_i$ has as its $(\xi_r\eta_i, \xi_s\eta_j)$ entry the value x^{ij}_{rs}. Now $\mathrm{Tr}_{F''/F}(\zeta) = \mu_{11} + \cdots + \mu_{mm}$ and $\mathrm{Tr}_{F'/F}(\mu_{ii}) = x^{ii}_{11} + \cdots + x^{ii}_{nn}$. So

$$\mathrm{Tr}_{F'/F}(\mathrm{Tr}_{F''/F'}(\zeta)) = \sum_{r=1}^{n}\sum_{i=1}^{m} x^{ii}_{rr} = \mathrm{Tr}_{F''/F}(\zeta).$$

(c) Choose $\xi_1, \ldots, \xi_n$ to be an integral basis of F' over P. Then if $\zeta \in F'_P$ it follows that $\zeta\xi_i \in F'_P$ and so all the entries of the multiplication matrix of ζ with respect to $\xi_1, \ldots, \xi_n$ lie in F_P. Hence its trace lies in F_P. □

13.13 Extending differentials

The trace can be extended to a linear map taking the set of relatively constant repartitions X'' of F' into the set of repartitions X of F by setting $(\mathrm{Tr}(\Gamma'))_P = \mathrm{Tr}(\Gamma'_{P'})$, where P' is any place of F' lying over P.

Let ω be a differential of F. We define a map on X'' which we denote by $\omega\,\mathrm{Tr}$ by putting

$$\omega\,\mathrm{Tr}(\Gamma') = \omega(\mathrm{Tr}(\Gamma')).$$

PROPOSITION *The map $\omega\,\mathrm{Tr}$ is a K-linear map from X'' to K that is zero on F' and on $X'([\omega]^{F'}) \cap X''$.*

Proof Linearity follows directly because the trace is F-linear and so certainly K-linear, while ω is K-linear by definition and the product of linear maps is linear.

A constant repartition $\Gamma' = \xi$ certainly lies in X'' and $\mathrm{Tr}(\xi)$ lies in F. Since ω is zero on F it follows that $\omega\,\mathrm{Tr}$ is zero on F'.

Suppose that $\Gamma' \in X''$ lies in $X'([\omega]^{F'})$, and let P be any place of F. Choose $x \in F$ such that $\nu_P(x) = [\omega](P)$. Then $\nu_{P'}(x\Gamma') \geq 0$ for any place P' of F' with $P' \cap F = P$. Therefore $x\Gamma'_{P'}$ lies in $F_{P'}$. Hence $\mathrm{Tr}(x\Gamma'_{P'}) = x\,\mathrm{Tr}(\Gamma'_{P'})$ lies in F_P and thus $\nu_P(\mathrm{Tr}\,\Gamma') + [\omega](P) \geq 0$. That shows that $\mathrm{Tr}(\Gamma')$ lies in $X([\omega])$ and so $\omega\,\mathrm{Tr}(\Gamma') = 0$. □

13.14 The genus of a separable extension

We can now estimate the behaviour of the genus under separable extensions.

THEOREM *Let F'/F be a finite separable extension of the function field F over K. Let the genera of F' and F be denoted by γ' and γ respectively and let the degrees of the fields of constants $\bar{K}'$ and $\bar{K}$ of F' and F over K be κ' and κ respectively. Then*

$$\gamma' - \kappa' \geq |F' : F|(\gamma - \kappa).$$

Proof Let ω be a non-zero differential of F. We extend the map $\omega\,\mathrm{Tr}$ from X'' to the whole of X' by putting $\omega\,\mathrm{Tr}(\Gamma') = 0$ for $\Gamma' \in X'([\omega]^{F'})$. That is consistent, since we showed in Proposition 13.13 that if $\Gamma' \in X'([\omega]^{F'}) \cap X''$, then $\omega\,\mathrm{Tr}(\Gamma')$ is already 0. That defines $\omega\,\mathrm{Tr}$ for all differentials because by Proposition 13.11 every repartition of F' is the sum of a relatively constant repartition and an element of $X'([\omega]^{F'})$.

The map $\omega\,\mathrm{Tr}$ is zero on F' and on $X'([\omega]^{F'})$ so it is a differential of F'. It remains to show that $\omega\,\mathrm{Tr}$ is not the zero map. By definition there exists a repartition Γ of F such that $\omega(\Gamma) \neq 0$ and by assumption there exists $\xi' \in F'$ such that $x = \mathrm{Tr}(\xi') \neq 0$; putting $\xi = \xi'/x$ we find that $\mathrm{Tr}(\xi) = 1$. Now we define the relatively constant repartition Γ' by putting $\Gamma'_{P'} = \xi\Gamma_P$ where $P = P' \cap F$. Then $\mathrm{Tr}(\Gamma'_P) = \Gamma_P\,\mathrm{Tr}(\xi) = \Gamma_P$ and so $\omega\,\mathrm{Tr}(\Gamma') = \omega(\Gamma) \neq 0$.

The degree of the divisor of $\omega\,\mathrm{Tr}$ is $2(\gamma' - \kappa')$ and we have seen that this divisor contains $[\omega]^{F'}$ because $\omega\,\mathrm{Tr}$ is zero on $X'([\omega]^{F'})$. The degree of $[\omega]$ is $2(\gamma - \kappa)$ and when a divisor is raised to F' its degree is multiplied by $|F' : F|$ (see Proposition 13.7). That proves that

$$2(\gamma' - \kappa') \geq |F' : F|\,2(\gamma - \kappa).$$

Dividing by 2 gives the statement of the theorem. □

13.15 The different

If we abuse notation and denote the differential of F' constructed from the non-zero differential ω in the proof of Theorem 13.14 by $\omega\,\mathrm{Tr}$, then the divisor

$[\omega\,\mathrm{Tr}] - [\omega]^{F'}$ is independent of ω. For any other non-zero differential of F has the form $x\omega$ with some non-zero $x \in F$. Then $[x\omega\,\mathrm{Tr}] = [\omega\,\mathrm{Tr}] + [x]^{F'}$ and $[x\omega] = [\omega] + [x]$. Hence

$$[x\omega\,\mathrm{Tr}] - [x\omega] = [\omega\,\mathrm{Tr}] + [x]^{F'} - [\omega]^{F'} - [x]^{F'} = [\omega\,\mathrm{Tr}] - [\omega]^{F'}.$$

That justifies the following definition.

DEFINITION Let F' be a finite separable extension of the function field F and let ω be any non-zero differential of F. The divisor $[\omega\,\mathrm{Tr}] - [\omega]^{F'}$ is called the *different* of F' over F.

Theorem 13.14 can now be restated in the following form.

THEOREM Hurwitz genus formula *Let D' be the different of the finite separable extension F' of the function field F. Then*

$$\gamma' - \kappa' = |F' : F|(\gamma - \kappa) + d(D')/2. \qquad \square$$

You should be aware that when Hurwitz proved his formula he was not using this definition of the different but one which is related to the ramification of the places of F in F'. That definition is discussed in some detail in Stichtenoth's book (1991) and also in Chevalley (1951). Tate (1952) showed that it is possible to use other maps than the trace, making it possible to investigate the inseparable case also. The generalized different is discussed in Stichtenoth's German lecture notes (1978).

13.16 Inseparable extensions

As an application of these ideas we shall prove Lüroth's theorem, which states that a subfield of a rational function field is either contained in the field of constants or a rational function field itself. The theorem can be proved by more direct algebraic means (see Cohn 1977). It is also slightly mysterious, because it is not true for fields of rational functions in more than two indeterminates. The proof given here will rely on showing that the genus of such a subfield is 0 and then finding a rational place.

For the proof we need to consider inseparable extensions. By definition the trace of an inseparable extension F' over F is 0. That applies also to the elements of F itself. For any basis, the matrix representing multiplication by an element $x \in F$ is just the diagonal matrix xI. This has trace $|F' : F|\,x$. So the degree of an inseparable extension must be a multiple of the characteristic of the base field. The following lemma strengthens this observation for purely inseparable extensions.

DEFINITION An extension field F' of field F is called *inseparable* if it is not separable. It is called *purely inseparable* if no intermediate field $F \subset F'' \subseteq F'$ is separable.

LEMMA *Let F' be a purely inseparable algebraic extension of the field F. Then the minimal polynomial of any element $\alpha \in F'$ has the form $x^{p^n} - a$ for some $a \in F$ and some power p^k of the characteristic of F.*

Proof The subfield $K(\alpha)$ has a basis $1, \alpha, \ldots, \alpha^n$ over F' for some n. With respect to this basis the multiplication matrix of α is

$$\begin{bmatrix} 0 & 0 & 0 & \cdots & -a_0 \\ 1 & 0 & 0 & \cdots & -a_1 \\ \vdots & \ddots & \ddots & & \vdots \\ \vdots & & \ddots & \ddots & \vdots \\ 0 & 0 & 0 & 1 & -a_{n-1} \end{bmatrix}$$

where all the entries are zero except for those just below the diagonal, which are ones, and the negative coefficients of the minimal polynomial

$$t(\dot{x}) = \dot{x}^n + a_{n-1}\dot{x}^{n-1} + \cdots + a_1\dot{x} + a_0$$

of α in the last column. This matrix has $t(\dot{x})$ as its characteristic polynomial. So its trace is the sum of the roots $\lambda_1, \ldots, \lambda_n$ of $t(x)$ (which are the eigenvalues of the matrix). Then the trace of $\dot{x}^k$ is the sum of the powers $\lambda_1^k, \ldots, \lambda_n^k$ (which are the eigenvalues of the multiplication matrix of x^k). Therefore since all traces are 0, we have

$$\begin{bmatrix} 1 & \cdots & 1 \\ \lambda_1 & \cdots & \lambda_n \\ \lambda_1^2 & \cdots & \lambda_n^2 \\ \vdots & \vdots & \vdots \\ \lambda_1^{n-1} & \cdots & \lambda_n^{n-1} \end{bmatrix} \begin{bmatrix} 1 \\ 1 \\ 1 \\ \vdots \\ 1 \end{bmatrix} = \begin{bmatrix} 0 \\ 0 \\ 0 \\ \vdots \\ 0 \end{bmatrix}.$$

It is well known that the determinant of the *Vandermonde* matrix on the left hand side is $\prod_{i<j}(\lambda_j - \lambda_i)$ (see my book 1996) so $\lambda_i = \lambda_j$ for some $i \neq j$. That implies that λ_i is a double root of $t(\dot{x})$ and so also a root of the derivative $t'(x)$. But $t(\dot{x})$ is irreducible and thus the minimal polynomial of λ_i. So it must divide $t'(\dot{x})$. That is only possible if $t'(\dot{x}) = 0$ which in turn is only possible if the characteristic of F is p and $t(\dot{x})$ is a polynomial in x^p, say $t(\dot{x}) = s(\dot{x}^p)$. Then $s(\dot{x})$ is the minimal polynomial of α^p over F and so $F'(\alpha^p)$ is a proper subfield of $F'(\alpha)$. The minimal polynomial of α over $F'(x^p)$ is $\dot{x}^p - \alpha^p$.

Repeating this argument we find that $x^q \in F$ for some $q = p^k$ and $t(\dot{x})$ is $\dot{x}^q - \alpha^q$ as claimed. □

13.17 Subfields of rational function fields

THEOREM Lüroth's theorem *Let K be a field and F be a subfield of the field of rational functions $K(x)$ strictly containing F. Then F is a field of rational functions.*

Proof Since F is not contained in the field of constants of $K(x)$, it contains a rational function $u = f(x)/g(x)$. Then $g(\dot{x})u - f(\dot{x})$ is a polynomial with coefficients in F and x as a root. So x is algebraic over F and therefore $K(x)$ is a finite extension of F. Any rational place of $K(x)$ intersects F in a place and since the residue field cannot get larger in the intersection and must contain K it follows that the intersection is also a rational place. So by Theorem 11.12, we need only show that the genus of F is 0.

If the trace of an extension F' over F is non-zero, then it maps F' onto the whole of F because it is F-linear. Hence it follows immediately from the trace formula (Proposition 13.12b) that a separable extension of a separable extension is separable. Therefore there is a maximal separable extension F' of F. As F' is certainly not contained in K, it is also a function field with K as its field of constants. Applying the Hurwitz genus formula to F and F' we obtain

$$\gamma' - 1 \geq |F' : F|\,(\gamma - 1).$$

Hence if we can show that $\gamma' = 0$ it follows that $\gamma = 0$.

By definition the extension $K(x)$ over F' is purely inseparable, and so by Lemma 13.16 the minimal polynomial of x over F' is $\dot{x}^q - x^q$ for some power q of the characteristic of K. Then $K(x^q) \subseteq F' \subset K(x)$ and $|K(x) : F'| = q = |K(x) : K(x^q)|$. Therefore $F' = K(x^q)$. That is the field of rational functions in x^q and so it has genus 0. □

13.18 Exercises

1. Show that the polynomial $x^3 + y^3 + z$ is absolutely irreducible over $K = \mathsf{F}_2(z)$ and that the curve C: $x^3 + y^3 + z = 0$ is smooth over K.

Show that the function field F of the curve C: $x^3 + y^3 + z = 0$ over K has K as its field of constants and genus 1.

(*This can be shown directly, as in Chapter* 4, *or using the Plücker formula.*)

Show that F also contains $K' = \mathsf{F}_2(x)$ and that F is isomorphic to $K'(y)$, which is a field of functions of genus 0 with field of constants K'.

2. With the notation of Paragraph 13.2, show that if $x \in F$, then $\pi(\omega x) = (\pi\omega)x$.

Show also that if $K \neq \bar{K}$ a non-zero K-differential is not a $\bar{K}$-differential.

3. Show that if F is a field of functions over K and F'/K' is a constant field extension, then F is a field of functions over K such that K' is contained in its field of constants.

4. Let F be the field of rational functions $K(x)$ over K. Let $F' = K(y)$ where $x = y^2(y+1)$. Show that there are two places of F' lying over the place P of F corresponding to x and that they both have relative degree 1.

Show further that the ramification indices of the places are 1 and 2.

In the remaining exercises F is a function field over K and F' is a finite algebraic extension of F.

5. Show that $P' \cap F$ is a place of F for any place P' of F'.

Show that for any place P of F there exists a place P' of F' such that $P' \cap F = P$.

6. Let $\eta_1, \ldots, \eta_n$ be a basis of F' over F and let P be a place of F. Show that there exists $\eta = \sum y_i\eta_i \in F'_P$ such that $y_1 \neq 0$.

7. Verify the claims made in the proof of Proposition 13.10 that the elements $\xi_1, \eta_2, \ldots, \eta_n$ form a basis of F' over F and that the final basis $\xi_1, \ldots, \xi_n$ is integral over P.
8. Show that the relatively constant repartitions of F' form a proper subspace of the set of all repartitions of F'.
9. Fill in the details of the proof of Proposition 13.12a, which states that the trace of ξ does not depend on the choice of basis used to calculate the multiplication matrix of ξ.
10. Show that in a finite field of characteristic p every element is a pth power. Using Lemma 13.16 deduce that a finite field has no purely inseparable finite extension and hence no inseparable finite extension. Fields with this property are called *perfect*.

14

CURVES AND FUNCTION FIELDS

In this chapter we consider function fields in relation to plane curves and develop those parts of the theory that are related to the equation of a curve and a particular coordinate system (which we shall always take to be the (x, y)-system).

14.1 Basic definitions

We shall use the definitions of curves and points that were introduced in Chapter 2.

DEFINITION Let $f(\dot{x}, \dot{y})$ be an absolutely irreducible polynomial in $K[\dot{x}, \dot{y}]$ of degree d and define

$$g(\dot{u}, \dot{v}) = \dot{u}^d f(1/\dot{u}, \dot{v}/\dot{u});$$
$$h(\dot{w}, \dot{z}) = \dot{w}^d f(\dot{z}/\dot{w}, 1/\dot{w}).$$

Then the algebraic curve C defined by $f(x, y)$ is the union of the three affine curves C_1: $f(x, y) = 0$, C_2: $g(u, v) = 0$, and C_3: $h(w, z) = 0$, points being allowed to have coordinates in any finite extension L of K. We call these three curves the *affine components* of C. The full curve is also denoted by C: $f(x, y) = 0$.

An *(x, y)-point* of the curve is a zero P: $(x{=}\alpha, y{=}\beta)$ of $f(x, y)$ with α and β in some finite extension field L of K. Its *degree* is $\deg(P) = |K[\alpha, \beta] : K|$. A point of degree 1 is called *rational* over K.

We retain the definition of the coordinate ring and function field from Chapter 3.

DEFINITION The ring $K[C] = K[x, y] = K[\dot{x}, \dot{y}]/f(\dot{x}, \dot{y})$ is called the *(x, y)-coordinate ring* of the curve C: $f(\dot{x}, \dot{y}) = 0$. Its field of fractions is called the *function field* $K(x, y) = K(C)$ of C.

It was shown in Chapter 3 that the function field $K(C)$ is independent of the coordinate system. We now verify that it really is a field of algebraic functions.

PROPOSITION *Let C: $f(x, y) = 0$ be an algebraic curve. Then the field $K(C)$ is a field of algebraic functions over K.*

Proof We assume without loss of generality that the polynomial $f(\dot{x}, \dot{y})$ involves $\dot{y}$ explicitly. In that case no polynomial in $K[\dot{x}]$ is a multiple of $f(\dot{x}, \dot{y})$. Therefore x is not algebraic over K. On the other hand y is a root of the polynomial $f(x, \dot{y})$ in $K(x)[\dot{y}]$ and so $K(C)$ is a finite extension of $K(x)$. Thus $K(C)$ satisfies the conditions for a field of algebraic functions. □

14.2 Points and places

For points we shall retain the definition of *local ring*, *x-order*, and *y-order* given in Chapter 3. The local ring $K[C]_P$ consists of those functions in $K(C)$ that can be written in the form $f(x,y)/g(x,y)$ with $g(P) \neq 0$. If the point is (α, β), then the x-order of φ is the highest power n such that $\varphi = (x-\alpha)^n \psi$ with $\psi \in K[C]_P$.

We shall use the word *place* in its formal algebraic sense as a special kind of subring of $K(C)$ and talk of a *conjugacy class* of points if we wish refer to a class of points at which all functions have the same value. Observe the difference in notation: $K[C]_P$ denotes the local ring of P, while $K(C)_P$ denotes the place ring of a place P. We can restate the definition of simplicity as follows.

PROPOSITION *A point is simple if and only if its local ring $K[C]_P$ is a place.*
Proof $K[C]_P$ is indeed a ring. For if $g_1(x,y)/h_1(x,y)$ and $g_2(x,y)/h_2(x,y)$ lie in $K[C]_P$, then their sum and product both have denominator $h_1(x,y)h_2(x,y)$ which is non-zero at P. So they both lie in $K[C]_P$.

$K[C]_P$ is not the whole of $K(C)$ for if the point P is P: $(x{=}\alpha, y{=}\beta)$, and the minimal polynomial of α over K is $t(x)$, then $1/t(x) \notin K[C]_P$.

The condition of simplicity is that for any rational function φ one of φ and $1/\varphi$ at least lies in $K[C]_P$. That is literally the condition that the ring $K[C]_P$ is a place. □

14.3 The field of constants of $K(C)$

A major task in this chapter will be to elucidate the connection between points of a (smooth) curve and places of its function field. For that purpose we need some purely algebraic results.

First we shall show that for a finite field K the field of constants of $K(C)$ is always K itself, which simplifies the application of the general theory of function fields.

PROPOSITION *Let C: $f(x,y)=0$ be an algebraic curve with function field $F = K(C)$. If K is finite, then the field of constants of F is K.*
Proof Suppose that $\varphi(x,y) \in K(C)$ is algebraic over K but not an element of K, and let $h(\dot{u})$ be the minimal polynomial of φ in $K[\dot{u}]$. Then $h(\varphi(x,y)) = 0$ so $h(\varphi(\dot{x},\dot{y})) \equiv 0 \pmod{f}$. Now let L be a finite splitting field of $h(\dot{u})$ over K. Then h splits into a product of distinct linear factors over L, so

$$h(\varphi(\dot{x},\dot{y})) = \prod(\varphi(\dot{x},\dot{y}) - \alpha_i).$$

As $f(\dot{x},\dot{y})$ is absolutely irreducible we must have $\varphi(\dot{x},\dot{y}) - \alpha \equiv 0 \pmod{f}$ with $\alpha = \alpha_i$ for some i. Thus $\varphi(x,y) = \alpha$ and by our hypothesis $\alpha \notin K$.

Let $|K| = q$. Then the map σ that takes β to β^q is an automorphism of L leaving elements of K fixed and moving all other elements of L. It follows that $f(\dot{x},\dot{y}) = \sigma f(\dot{x},\dot{y})$, and $\sigma\varphi(\dot{x},\dot{y}) = \varphi(\dot{x},\dot{y})$. Hence $\varphi(\dot{x},\dot{y}) - \alpha^q \equiv 0$ (mod f). Taking the difference implies that over L, f divides $\alpha - \alpha^q$, which is a non-zero constant. That is impossible for a polynomial of positive degree. So the assumption that $\varphi(x, y)$ is algebraic over K is not tenable. □

Remark Using standard field theory this proof can be adapted to any *perfect* field (see Exercise 1). In particular, the assumption that K is finite can be replaced by the assumption that it has characteristic 0.

14.4 Algebraic extensions

We require also a purely algebraic result that can easily be derived from the properties of algebraic function fields. It extends that standard field theoretical result that $K[x]$ is a field if and only if x is algebraic over K.

PROPOSITION *Let K be a field. Then $K[x, y]$ is a field if and only if x and y are both algebraic over K.*

Proof If x is algebraic over K then $K[x]$ is a field, and if y is algebraic over $K[x]$ then $K[x, y]$ is a field.

Assume conversely that $K[x, y]$ is a field. Let $L = K(x)$ be the field of fractions of $K[x]$ contained in $K[x, y]$. Then $K[x, y] = L[y]$ and so y is algebraic over L. Hence $|K[x, y] : L|$ is finite and it follows that, if x is not algebraic over K, then $K[x, y]$ is a function field.

Let P be any place. If $\nu_P(x) \geq 0$ and $\nu_P(y) \geq 0$, then $\nu_P(f(x, y)) \geq 0$ for all polynomials $f(x, y) \in K[x, y]$. Thus $K[x, y]$ must be contained in the place ring of P. Since $K[x, y]$ is assumed to be the complete function field that is impossible. So we have $\nu_P(x) < 0$ or $\nu_P(y) < 0$ for every place P. By the zeros theorem there are only finitely many such places, but by the existence theorem any function field has infinitely many places. Thus the assumption that x is not algebraic over K leads to a contradiction. By symmetry it also follows that y is algebraic over K and our proposition is established. □

14.5 Points and ideals

As a consequence of the proposition just proved we can identify (conjugacy classes of) (x, y)-points of a curve C with *maximal* ideals of its coordinate ring $K[C]$ (it is a standard convention to use the term 'maximal ideal' for 'maximal proper ideal'). Recall that the ideal I of the commutative ring is maximal if and only if the factor ring R/I is a field.

Every (x, y)-point P: $(x{=}\alpha, y{=}\beta)$ defines a homomorphism from the coordinate ring $K[x, y]$ onto the field $K[\alpha, \beta]$ taking $g(x, y)$ to $g(\alpha, \beta)$. The kernel of this evaluation map must be a maximal ideal of $K[x, y]$. Suppose conversely we are given a homomorphism σ of $K[x, y]$ onto a field L. It maps, say, x to α and y to β. Then $L = K[\alpha, \beta]$ and so by Proposition 14.4, α and β are algebraic, and σ is the evaluation map of the point P: $(x{=}\alpha, y{=}\beta)$.

In Chapter 4 we defined two (x, y)-points P and Q to be conjugate if they were zeros of the same set of polynomials in $K[x, y]$ and we have just seen that this set must be a maximal ideal. We formalize this discussion in the following proposition.

PROPOSITION *The map taking every point P of a curve to the ideal of functions that evaluate to 0 at P induces a bijection from conjugacy classes of (x, y)-points to maximal ideals of $K[x, y]$.* □

14.6 Places over points

The existence theorem for function fields (Theorem 9.7) tells us that if we choose a non-zero prime ideal I of $K[C] = K[x, y]$, then there exists a place P such that $P \cap K[C] = I$. We shall use this theorem to show that all non-zero prime ideals of $K[C]$ are maximal and to establish a link between conjugacy classes of (x, y)-points and certain places.

DEFINITION A place P of $K(C)$ is called *finite* if $K[C] \subseteq K(C)_P$. Otherwise it is called *infinite*. If $P \cap K[C]$ is a maximal ideal M corresponding to a point Q, then P is said to lie *over* M or Q.

The definition of finite depends on the coordinate system being used. So it would be more precise to say '(x, y)-finite', '(u, v)-finite', or '(w, z)-finite'; but in this chapter we shall only use the term with reference to the (x, y)-system—except in the proposition below, which shows that every place is finite for one of the three coordinate systems and that every finite place lies over a maximal ideal of $K[C] = K[x, y]$.

PROPOSITION *Let P be a place; then P is finite for one at least of the three coordinate systems (x, y), (u, v), and (w, z). If P is (x, y)-finite, then $P \cap K[C]$ is maximal.*

Proof To show that P is finite for one of the systems we must establish that its ring contains both x and y, or both u and v, or both w and z. The ring must contain at least one member of each of the inverse pairs (x, v), (y, w), and (u, z). The only two combinations that do not contain a coordinate pair are x, w, u and y, v, z. These triples multiply to 1. So if $K(C)_P$ contains all three members of one of them, they are invertible in $K(C)_P$. Then the ring contains all six functions and is finite in all three systems.

Now suppose that P is (x, y)-finite. Note that $P \cap K[C] \neq 0$. For in that case every non-zero element of $K[C]$ would be invertible in $K(C)_P$ and hence $K(C)_P$ would contain the whole of $K(C)$.

The factor ring of $K[C]/(P \cap K[C])$ is a subring of the field $L = K(C)_P/P$. By Proposition 11.1, $K(C)_P/P$ is a finite extension of K. If the images of x and y in $K(C)_P/P$ are α and β, then $K[C]/(P \cap K[C]) = K[\alpha, \beta]$. Furthermore, since α and β are algebraic over K, it is a field. Hence $P \cap K[C]$ is maximal. □

By the existence theorem (Theorem 9.7), every prime ideal I of $K[C]$ extends to a place P such that $P \cap K[C] = I$. That gives the following corollary.

COROLLARY (a) *All non-zero prime ideals of $K[C]$ are maximal.* □

Applying the proposition to all three coordinate systems we obtain the following further corollaries relating points and places.

COROLLARIES (b) *To every point Q there is a place lying over Q.*
(c) *Every place lies over a unique conjugacy class of points.* □

If a maximal ideal of $K[C]$ corresponds to a class of simple points, then their local ring is already a place, and therefore there is just one place over the class. For singular points the place is sometimes uniquely determined and sometimes not.

EXAMPLES Consider the curve $y^2 = x^3$ over R and the point P: $(0,0)$. This curve has a *cusp* at P with a unique tangent $y = 0$. Algebraically this is reflected in the following facts.

(i) P is singular.
(ii) Since $(y/x)^2 = x$, the fraction y/x must lie in the place ideal of any place extending $K[C]_P$.
(iii) $K[C]_P$ extends to a unique place $K(C)_P$ with $\nu_P(y/x) = 1$.
You can see this by using the fact that $(y/x)^2 = x$ and $(y/x)^3 = x(y/x) = y$. Substitute these expressions to write any rational function $\varphi \in K(C)$ in the form $g(y/x)/h(y/x)$ where g and h are polynomials in y/x. Cancelling common factors y/x we can assume that one of g and h has a non-zero constant term. $K(C)_P$ consists of those rational functions for which h has a non-zero constant term.
(iv) This place is *ramified* over $K(x)$ and $K(y)$ because no element φ in either field has value $\nu_P(\varphi) = 1$. The *ramification index* over $K(x)$ is 2 because $\nu_P(x) = 2$; over $K(y)$ the ramification index is 3.

Now consider the α-shaped curve $y^2 = x^3 + x$ and the point P: $(0,0)$. In characteristic different from 2 this is a *double* point having two tangents with slopes ± 1. It is difficult to draw pictures in characteristic 2, but the fact that we have a 'double' tangent and the calculations below suggest that in characteristic 2 the point should be considered as a cusp. Algebraically, these statements are reflected in the following facts.

(i) P is singular.
(ii) Since $(y/x)^2 = x + 1$, the fraction y/x must lie in the place ring of any place extending $K[C]_P$ but it must lie outside the place ideal. Its value at P must be ± 1.
(iii) There are two places extending $K[C]_P$, one in which $y/x + 1$ lies in the place ideal, and the other in which $y/x - 1$ lies in the place ideal. In characteristic different from 2, these two places are different, as 2 cannot lie in a place ideal. We consider the first of these, denoting it by P'.
(iv) The equations

$$(y/x + 1)^2 - 2(y/x + 1) = y^2/x^2 - 1 = x$$

$$(y/x+1)^3 - 3(y/x+1)^2 + 2(y/x+1) = (y/x+1)x - x = y$$

make it possible for us to convert any rational function $\varphi \in K(C)$ into the form $g(y/x+1)/h(y/x+1)$. As before the place ring consists of those functions for which h has a non-zero constant term.

(v) In characteristic different from 2, there is no ramification over $K(x)$ or $K(y)$ as our equations show that $\nu_{P'}(x) = 1 = \nu_{P'}(y)$. In characteristic 2 we have ramification as the terms in our equations with even coefficients vanish and then both x and y have order 2.

The examples suggest that there is a link between the ramification indices and the number of places extending the local ring of a singular point similar to Proposition 13.6. That is indeed true. Fulton (1989) is a good reference for this topic.

14.7 The zeros theorem and singular points

The zeros theorem for curves is now an easy consequence of the zeros theorem for function fields.

THEOREM Zeros theorem *Let C: $f(x, y) = 0$ be an affine curve over K and let $g(x, y)$ be a non-constant polynomial in $K[x, y]$. Then there exists a point P: $(x{=}\alpha, y{=}\beta)$ of C with α, β in some finite extension field of K, such that $g(\alpha, \beta) = 0$. The number of such points is finite.*

Proof By the zeros theorem for function fields, g lies in finitely many places (which we equate as usual with their place ideals). As g is a polynomial, all these places are finite. The points at which g is zero are precisely those underlying these places. As each place determines a unique conjugacy class of points and each class has only finitely many members, the number of such points is finite. □

COROLLARY *Any curve C has only finitely many singular points.*

Proof In Theorem 3.12 it was established that a necessary condition for a point P: $(x{=}\alpha, y{=}\beta)$ of the curve C: $f(x, y) = 0$ to be singular is that P is a zero of $g = \partial f/\partial x$. By the theorem, there are only finitely many zeros of g. Among these lie all the singular points of C. □

14.8 Simple points

In Chapter 3 we defined a curve to be *smooth* if all its points are simple, and *(x, y)-smooth* if all its (x, y)-points are simple. Most of the theory of Chapter 4 and Goppa codes was developed for smooth curves.

From a purely formal point of view, the requirement that a curve should be smooth is unnecessary for geometric codes, as the entire theory can be phrased in terms of places. However, smoothness allows us to identify conjugacy classes of points with places. Otherwise, some points will correspond to multiple places or there may be ramification. Smoothness also eases calculations, because of the horizon theorem, which we shall prove in the next paragraph.

Identifying points with places makes the degree theorem for smooth curves into a special case of the degree theorem for function fields. The only work we need to do is to prove that for a simple point the order function of the point and the valuation of the corresponding place coincide.

PROPOSITION *Let P be a simple point of the curve C and denote by μ the order function of P and by ν the valuation of the corresponding place. Then $\mu = \nu$.*

Proof In the proof of Theorem 10.5 the function value $\nu(g)$ was defined as the greatest length of a sequence $g = h_0, \ldots, h_n$ with $h_i \in K(C)_P$ and h_i/h_{i+1} in the maximal ideal of $K(C)_P$, or in our context, $h_i/h_{i+1}(P) = 0$.

Suppose that the point is $P\colon (\alpha, \beta)$. The order function μ is either the x-order or the y-order; say μ is the x-order. Then $\mu(g)$ is the highest power n such that $g = (x - \alpha)^n \varphi$ for some $\varphi \in K[C]_P = K(C)_P$. Since P is simple, this order is *good*, which means that $\mu(g) = 0$ implies that $g(P) \neq 0$.

Suppose that $\mu(g) = m$ and $g = (x - \alpha)^m \varphi$ with $\varphi \in K(C)_P$. Then the sequence $h_k = (x - \alpha)^{m-k} \varphi$ satisfies the conditions for ν and so $\nu(g) \geq m$.

On the other hand, let $\nu(g) = n$ and choose a sequence $g = h_0, \ldots, h_n$ with $h_i \in K(C)_P$ and $h_i/h_{i+1}(P) = 0$. The assumption that the x-order of P is good implies that $h_i/h_{i+1} = (x - \alpha)\varphi_i$ with $\varphi_i \in K[C]_P$ for each i. Then

$$g = h_n \prod_{i=0}^{n-1} \frac{h_i}{h_{i+1}} = (x - \alpha)^n h_n \prod_{i=0}^{n-1} \varphi_i.$$

This is a representation $g = (x - \alpha)^n \psi$ with $\psi \in K[C]_P$ and so $\mu(g) \geq n$.

Thus $\mu(g) = \nu(g)$ as claimed. □

The proof has established that $(x - \alpha)$ is a *local parameter* for P.

COROLLARY Degree theorem *Let $C\colon f(x, y) = 0$ be a smooth curve defined over K, and let $\varphi \in K(C)$. Then $\nu_P(\varphi) \neq 0$ for finitely many point classes P, and summing over all point classes we have*

$$\sum \nu_P(\varphi) d(P) = 0.$$

Proof Since all points are simple, places and point classes are in one-to-one correspondence, and the order functions of the points are just the valuations of the places. Therefore this is just a restatement of the degree theorem for function fields. □

14.9 The horizon theorem

The remaining unproved theorems from Chapter 4 are the horizon theorem and the Plücker formula for the genus. These cannot be derived as immediate consequences of results we have proved for function fields, because they involve concepts that have no direct equivalents in the theory we have built so far. For the horizon theorem the critical concept is (x, y)-smoothness. Here is an algebraic characterization of that concept, from which the horizon theorem follows easily.

THEOREM *An affine curve C: $f(x,y)=0$ is (x,y)-smooth if and only if $K[C]$ is the intersection of all finite place rings.*

Proof Since we shall consider only (x,y)-points, we shall identify them with the maximal ideals of $K[C]$. A polynomial $h \in K[C]$ is zero at the point corresponding to a maximal ideal M, if and only if $h \in M$. So the local ring of the point $K[C]_M$ consists of all rational functions that can be written in the form g/h where g and h lie in $K[C]$ and $h \notin M$. This local ring is a place if and only if the point is simple (Proposition 14.2).

The first statement implies the second. The hypothesis of (x,y)-smoothness implies that the rings $K[C]_M$ for maximal ideals M of $K[C]$ form the complete set of finite place rings. Thus we need only show that $K[C]$ is the intersection of these rings. With the appropriate definitions that is true for any domain.

Let $\varphi \in K(C)$ and define $I_\varphi = \{h \in K[C] : h\varphi \in K[C]\}$. It is immediate that I_φ is an ideal of $K[C]$, because if $h\varphi \in K[C]$, $k\varphi \in K[C]$, and $g \in K[C]$, then $(gh+k)\varphi \in K[C]$ (and certainly $0 \in I_\varphi$).

Furthermore, $\varphi \in K[C]$ if and only if $1 \in I_\varphi$. Thus if $\varphi \notin K[C]$, then I_φ is a proper ideal of $K[C]$. Let M be a maximal ideal containing I_φ. Then $\varphi \notin K[C]_M$, because that would imply $\varphi = g/h$ with $g \in K[C]$ and $h \in K[C] \setminus M$. Hence $h \in I_\varphi \setminus M$ contradicting the choice of M.

That shows that if $\varphi \notin K[C]$ there is some M for which $\varphi \notin K[C]_M$. So the intersection of all $K[C]_M$ is a subset of $K[C]$. Of course, $K[C] \subseteq K[C]_M$ for any maximal ideal and so the intersection must be the whole of $K[C]$.

The second statement implies the first. Choose a maximal ideal M. We shall apply the strong approximation theorem to establish that M is contained in a unique place P. Then we shall show that $K(C)_P = K[C]_M$.

So assume that M is contained in at least two distinct places P and Q. Then for any element h of $K[C]$, $\nu_P(h) = 0$ if and only if $\nu_Q(h) = 0$, because each of these statements is equivalent to $h \in M$.

Apply the corollary to the strong approximation theorem with P and Q as our set of places, taking a place at infinity as the exceptional place, and choosing the values 1 and 0. The corollary implies that there exists φ with $\nu_P(\varphi) = 1$, $\nu_Q(\varphi) = 0$, and $\nu_R(\varphi) \geq 0$ for all finite places R. As $\nu_P(\varphi) \neq 0$ and $\nu_Q(\varphi) = 0$ it follows that $\varphi \notin K[C]$. On the other hand, $\nu_R(\varphi) \geq 0$ for all finite places, including P and Q. That contradicts the hypothesis that $K[C]$ is the intersection of all finite place rings. So every maximal ideal must be contained in a unique place.

Let P be the unique place extending M and let $\varphi \in K(C)_P$. We wish to show that $\varphi \in K[C]_M$. There is no place extending M and containing φ^{-1} in its place ideal. Therefore by the existence theorem, φ^{-1} and M cannot generate a proper ideal in $K[C][\varphi^{-1}]$. Thus there is an equation

$$1 = \sum_{i=0}^{n} g_i \varphi^{-i},$$

with $g_i \in M$ for $i = 0, \ldots, n$. We multiply this equation by φ^n and move the term for $i = 0$ to the left hand side, obtaining

$$(1 - g_0)\varphi^n = \sum_{i=1}^{n} g_i \varphi^{n-i}. \qquad (*)$$

We shall show that $(1 - g_0)\varphi \in K[C]$. Then since $1 - g_0 \notin M$ it follows that $\varphi = (1 - g_0)\varphi/(1 - g_0) \in K[C]_M$.

Multiplying eqn $(*)$ by $(1 - g_0)^{n-1}$ and putting $\psi = (1 - g_0)\varphi$ we obtain

$$((1 - g_0)\varphi)^n = \psi^n = \sum_{i=1}^{n}(1 - g_0)^{i-1} g_i((1 - g_0)\varphi)^{n-i} = \sum_{i=1}^{n} h_i \psi^{n-i}, \qquad (**)$$

where $h_i = (1 - g_0)^{i-1} g_i \in K[C]$.

Let Q be any finite place ring. Then $\nu_Q(h_i) \geq 0$ for all $i = 1, \ldots, n$. If $\nu_Q(\psi) < 0$, then $\nu_Q(h_i \psi^{n-i}) > \nu_P(\psi^n)$ for all $i = 1, \ldots, n$. That contradicts the fact that by eqn $(**)$,

$$\nu_Q(\psi^n) \geq \min\{\nu_Q(h_i \psi^{n-i}) : i = 1, \ldots, n\}.$$

Thus $\nu_Q(\psi) \geq 0$ for all finite places Q. Hence ψ is in the intersection of all their place rings $K(C)_Q$, which is $K[C]$ by hypothesis. So we have shown that for any element $\varphi \in K(C)_P$ there exists $g_0 \in M$ such that $(1 - g_0)\varphi \in K[C]$. Therefore $K(C)_P \subseteq K[C]_M$ and as the opposite inclusion holds by definition of P the two rings are equal. Thus by Proposition 14.2 the point corresponding to M is simple. □

COROLLARIES (a) Horizon theorem *A curve C: $f(x, y) = 0$ is (x, y)-smooth if and only if every element of $K(C)$ with no (x, y)-poles lies in $K[C]$.*
(b) *If C: $f(x, y) = 0$ is (x, y)-smooth and $g, h \in K[C]$ have the property that for every place P with $\nu_P(h) > 0$ we have $\nu_P(g) \geq \nu_P(h)$, then $g = hk$ with $k \in K[C]$.*

Proof (a) The condition that φ has no (x, y)-poles is merely a restatement of the property that $\varphi \in \bigcap K(C)_P$, where the intersection runs over all places whose rings containing $K[C]$.

(b) The poles of $k = g/h$ lie among the poles of g and the zeros of h. The condition ensures that none of the zeros of h is a pole of k. Hence all the poles of k are poles of g and are therefore infinite. By the horizon theorem it follows that $k \in K[C]$. □

14.10 Extending the base field

The proof of the Plücker formula that I shall present relies on extending the base field suitably. So in this paragraph we shall consider the effect of replacing the field K over which the curve C is defined by a finite extension K'.

By definition the coordinates of the points of C are allowed to lie in extensions of the base field. So the points of C do not change. Let P: $(x{=}\beta, y{=}\gamma)$ be a point of C. It is possible that the minimal polynomial of β over K is no longer irreducible over K'. So a K-conjugacy class of points of C may split into several K'-conjugacy classes.

We make a simplifying assumption that corresponds to the one made in Paragraph 13.3.

ASSUMPTION The curve C has a simple K-rational point.

The point remains simple and rational over every finite extension K' of K, and so by Corollary 11.1, K' is always the field of constants of $K'(C)$ and $\kappa = 1$.

It then follows from Proposition 13.4 that the genus of $K'(C)$ is the same as that of $K(C)$.

14.11 Plücker's formula for the genus

The proof of the Plücker formula given here falls into two sections. First the formula is derived as an inequality that holds for all curves. Then it is shown that the inequality is an equation for smooth curves.

THEOREM Plücker inequality *Let K be a field and let C: $f(x, y) = 0$ be a curve defined over K with a simple rational point. If the degree of the absolutely irreducible polynomial $f(\dot{x}, \dot{y})$ defining C is n, then the genus γ of C satisfies $\gamma \leq (n-1)(n-2)/2$.*

Proof Let $f(\dot{x}, \dot{y}) = \sum_{i,j} a_{ij}\dot{x}^i\dot{y}^j$ have degree n. We first assume that the coefficient $a_{0n} \neq 0$. In that case we can write

$$f(\dot{x}, \dot{y}) = \sum_{i=0}^{n} f_i(\dot{x})\dot{y}^{n-i},$$

where $\deg f_i \leq i$ for all i and, in particular, $f_0 = a_{0n}$, which is a non-zero constant. It follows that $|K(C) : K(x)| = n$.

Let $D = [x]_\infty$ be the divisor of poles of x. The zeros theorem tells us that $d(D) = n$. We shall estimate $l(mD)$ and then derive the estimate for the genus from Riemann's theorem.

We first show that $y \in L(D)$, that is $\nu_P(y) \geq \min\{\nu_P(x), 0\}$ for all places P. For suppose that $\nu_P(y) < \min\{\nu_P(x), 0\}$ for some P. If $\nu_P(x) \geq 0$, then $\nu_P(f_i(x)) \geq 0$. Hence $\nu_P(y) < 0$ implies that for all $i = 1, \ldots, n$,

$$\nu_P(a_{0n}y^n) = \nu_P(y^n) < \nu_P(f_i(x))y^{n-i}.$$

If, on the other hand, $\nu_P(x) < 0$, then $\nu_P(f_i(x)) = \nu_P(x) \deg f_i$. Thus $\nu_P(y) < \nu_P(x)$ implies that for all $i = 1, \ldots, n$,

$$\nu_P(a_{0n}y^n) = \nu_P(y^n) < \nu_P(f_i(x))y^{n-i} = \nu_P(y^{n-i})\nu_P(x) \deg f_i.$$

In either case we find that $\nu_P(f(x, y)) = \nu_P(a_{0n}y^n) < 0$ contradicting the fact that $f(x, y) = 0$.

That confirms that $y \in L(D)$. Of course $x \in L(D)$ also. So for any value $m \geq 0$ the set $B_m = \{x^i y^j\}$ with $1 \leq j \leq n-1$ and $0 \leq i \leq m-j$ lies in $L(mD)$. Since $f(x,y)$ is the minimal polynomial of y over $K(x)$ no non-trivial linear combination of the elements of B_m can be 0, as that would provide a lower degree polynomial with y as a root. Therefore for $m \geq n$

$$\begin{aligned} l(mD) \geq \sum_{j=0}^{n-1}(m+1-j) &= n(m+1) - n(n-1)/2 \\ &= mn + 1 + (n-1) - n(n-1)/2 \\ &= mn + 1 - (n-2)(n-1)/2. \end{aligned}$$

By the degree theorem we have $\deg mD = mn$. Hence it follows from Riemann's theorem that $\gamma \leq (n-2)(n-1)/2$.

We now show that we can always assume that our equation $f(\dot{x},\dot{y})$ has $a_{0n} \neq 0$. To do that we replace $\dot{x}$ by $\dot{x}' = \dot{x} - b\dot{y}$. Then $f(\dot{x},\dot{y}) = f(\dot{x}' + b\dot{y}, \dot{y})$. The coefficient of $\dot{y}^n$ in this polynomial is $\sum_{i+j=n} a_{ij} b^i$. So provided that b is not a root of the non-zero polynomial $\sum_{i+j=n} a_{ij} \dot{u}^i$, the transformation will convert $f(\dot{x},\dot{y})$ into the required form.

We can always find a finite extension K' of K that contains a non-root of this polynomial. The genus γ of $K'(C)$ over the constant field K' satisfies the Plücker inequality. By Proposition 13.3 the genus of C over K' is the same as that over C and so the inequality holds in K. □

14.12 Plücker's formula is exact for smooth curves

It is possible to extend the Plücker inequality to an equation in which the additional terms depend on the nature of the singularities of the curve. Here we shall only show that if the curve is smooth, then the inequality becomes an equation.

THEOREM Plücker equation *Let K be a field and let C: $f(x,y) = 0$ be a smooth curve defined over K with a simple rational point. If the degree of the absolutely irreducible polynomial $f(\dot{x},\dot{y})$ is n, then the genus γ of C satisfies $\gamma = (n-1)(n-2)/2$.*

Proof (1) We adopt the notation used in the proof of the Plücker inequality. Using the second part of that proof we may assume that $f(x,y) = \sum_{i+j\leq n} a_{ij}x^i y^j$ and that $a_{0n} \neq 0$. Again we let $D = [x]_\infty$ and we know that $B_m = \{x^i y^j\}$ with $1 \leq j \leq n-1$ and $0 \leq i \leq m-j$ is a linearly independent subset of $L(mD)$. If we can show that B_m is a basis of $L(mD)$ then for $m \geq n$ it follows that

$$\begin{aligned} l(mD) = \sum_{j=0}^{n-1}(m+1-j) &= n(m+1) - n(n-1)/2 \\ &= mn + 1 - (n-2)(n-1)/2. \end{aligned}$$

Now the fact that $d(mD) = mn$ implies that $\gamma = (n-2)(n-1)/2$. So our task is to show that the set B_m spans $L(mD)$. In other words, every element of $L(mD)$ can be written as a linear combination of elements in B_m with coefficients in K.

(2) By the horizon theorem applied to the (x,y)-system, any element of $L(mD)$ can be represented as a polynomial in x and y. We can use the equation of the curve to eliminate all powers y^r with $r \geq n$ and then split the polynomial into its homogeneous components. We then have an element of the form $h(x,y) = \sum_r h_r(x,y)$ in $L(mD)$ where

$$h_r(x,y) = \sum b_{ij}x^i y^j \quad \text{with } 0 \leq i,j,\ i+j=r, \text{ and } j<n.$$

From the proof of the Plücker inequality it follows immediately that $h_r \in L(rD)$. Suppose we can show that $h_r \in L((r-1)D)$ occurs only for $h_r = 0$. Then letting $s = \deg h$, it follows that $h_s \notin L((s-1)D)$, while $h_r \in L((s-1)D)$ for all $r < s$. Therefore $h \in L(sD) \setminus L((s-1)D)$ and so $h \in L(mD)$ implies that $\deg h \leq m$. Thus B_m spans $L(mD)$ as required.

(3) It remains to show that $h_r \in L((r-1)D)$ implies that $h_r = 0$. We transform to the (u,v)-coordinate system. In that system we have $x = 1/v$ and $y = u/v$. So

$$h_r = v^{-r} \sum b_{ij} u^j = v^{-r} k(u),$$

where $k(\dot{u})$ is a polynomial of degree at most $n-1$. Assuming that $h_r \in L((r-1)D)$, it follows that for any place P,

$$-r\nu_P(v) + \nu_P(k(u)) + (r-1)D(P) \geq 0.$$

As D is the divisor of zeros of $v = 1/x$, it follows that for any zero P of v we have $D(P) = \nu_P(v)$, and so $\nu_P(k(u)) \geq \nu_P(v)$. The corollary to the horizon theorem (applied to the (u,v)-coordinate system) then gives

$$k(u) = vk_1(u,v) \qquad (*)$$

with $k_1(u,v) \in K[u,v]$.

(4) The equation of C in the (u,v)-system is $g(u,v) = 0$ where

$$g(\dot{u},\dot{v}) = \dot{v}^n f(1/\dot{v}, \dot{u}/\dot{v}) = \sum_{i+j\leq n} a_{ij}\dot{u}^j \dot{v}^{n-(i+j)} = g_1(\dot{u}) + \dot{v}g_2(\dot{u},\dot{v}).$$

The assumption that $a_{0n} \neq 0$ implies that $\deg g_1 = n$. Lifting eqn $(*)$ to the polynomial ring $K[\dot{u},\dot{v}]$ we obtain

$$k(\dot{u}) - \dot{v}k_1(\dot{u},\dot{v}) = k_2(\dot{u},\dot{v})\big(g_1(\dot{u}) + \dot{v}g_2(\dot{u},\dot{v})\big).$$

Mapping $\dot{v}$ to 0 we get $k(\dot{u}) = k_2(\dot{u},0)g_1(\dot{u})$, but $\deg k(\dot{u}) < n = \deg g_1(\dot{u})$. Hence $k(\dot{u}) = 0$ and so $h_r = v^{-r}k(u) = 0$, concluding the proof. □

Remark In the second step the proof requires that C is (x,y)-smooth, and in the third it requires that C is (u,v)-smooth, but it does not appear to require (w,z)-smoothness. In Exercise 6 you will be asked to explain this apparent asymmetry.

14.13 Exercises

1. Show that the statement of Proposition 14.3, which states that K is the constant field of $K(C)$, holds for any field K which has the property that every irreducible polynomial over K splits into distinct linear factors over a finite extension of K (such fields are called *perfect*).

You will need the following fact from general field theory.

If the irreducible polynomial $f(x)$ has two distinct roots α and β in an extension field L of K, then there is an automorphism of L fixing all the elements of K and taking α to β.

2. Show that any field of characteristic 0 is perfect and that a field of characteristic $p \neq 0$ is perfect if and only if every element in K has a pth root. Deduce that all finite fields are perfect (see also Exercise 10 of Chapter 13).

3. Extend the proof of Proposition 14.4 to an arbitrary finite number of elements: the ring $K[x_1, \ldots, x_n]$ is a field if and only if x_i is algebraic over K for all $i = 1, \ldots, n$.

4. Prove the corollaries of Paragraph 14.6:
   (a) All non-zero prime ideals of $K[C]$ are maximal.
   (b) To every (x, y)-point Q there is a place lying over Q.
   (c) Every place lies over a unique conjugacy class of points.

5. Let C be a plane curve defined over a field K and let K' be a finite extension of K. Show that if $K' = K[\alpha_1, \ldots, \alpha_n]$, then $K'[C] = K[C][\alpha_1, \ldots, \alpha_n]$.

Using the fact that α_i is algebraic over K and hence also algebraic over $K(C)$ deduce that $K'(C) = K(C)[\alpha_1, \ldots, \alpha_n]$.

Suppose now that $\alpha_1, \ldots, \alpha_n$ form a vector space basis of K' over K. Show that every monomial $\alpha_1^{r_1} \ldots \alpha_n^{r_n}$ can be written as a sum $s_1\alpha_1 + \cdots + s_n\alpha_n$ and deduce that the elements $\alpha_1, \ldots, \alpha_n$ span $K'(C)$ over $K(C)$ (that means that every element of K'(C) can be written as a sum $\varphi_1\alpha_1 + \cdots + \varphi_n\alpha_n$ with $\varphi_i \in K(C)$).

6. The proof of the Plücker equation explicitly uses (x, y)-smoothness and (u, v)-smoothness, but it apparently does not use (w, z)-smoothness. Explain why, nevertheless, the proof requires all the points of the curve to be simple.

15

MORE ON GOPPA CODES

In this final chapter we shall analyse the theoretical properties of geometric Goppa codes more deeply. Their main promise lies in providing codes with very high correction capabilities and reasonable decoding algorithms. Decoding algorithms were discussed in Chapters 6 and 7. Though these algorithms themselves are still too slow for practical use, refinements that improve their speed are already known (Justesen *et al.* 1992, Sakata *et al.* 1995). On the other hand we have not so far discussed how to construct good geometric Goppa codes, nor indeed what the word 'good' in this context means. So the chapter starts with a discussion of the Gilbert bound (also called the Gilbert–Varshamov bound (Gilbert 1952, Varshamov 1957) adapted from my book (1996). Most bounds on codes are upper bounds, stating that the parameters of a code must stay within certain limits, but the Gilbert bound is a lower bound. It states that codes with parameters satisfying certain bounds exist. We shall interpret the word 'good' to mean that the parameters of a code meet these bounds.

After introducing the Gilbert bound the chapter discusses what properties a curve needs in order that it can be used to construct a good Goppa code. It will emerge that what is needed is that the curve has a large number of rational places in relation to its genus. The first consequence of this is that it is not possible to restrict the investigation to smooth plane curves, since the number of rational points of a smooth plane curve is bounded by $q^2 + q + 1$ if it is defined over a field of order q. Curves in higher dimensional spaces are not defined by single equations and the classical theory becomes more complicated as the dimension increases. However, the theory of function fields is independent of the dimension and so it is simpler to use that theory.

There is still a price to pay. The determination of the genus becomes a formidable problem. There are some genus formulae (like the Hurwitz formula 13.15) which help, but the exact determination of the genus of a function field or high dimensional curve is often extremely difficult. Nevertheless it is possible to find curves that yield good codes. The chapter discusses two examples: the modular curves of Tsfasman *et al.* (1982), and the more explicit curves of Garcia and Stichtenoth (1995a, 1995b, 1996).

However, even the most concrete constructions of good geometric codes of large block length give only bounds on the genus. That leaves a major gap in the transition from theory to practice. It makes it troublesome to determine the rank, and very hard indeed to determine the minimum distance. The construction of a practical error processor without knowledge of the genus is currently not possible at all.

15.1 The volume of a ball of radius r

As with most estimates for codes, the Gilbert bound is based on counting the number of words at a given distance from a fixed word. Recall that the (Hamming) distance $d(u, v)$ between two words of length n is the number of entries i for which $u_i \neq v_i$.

DEFINITION Let K be a field of order q. The *ball* $B_q(u, r)$ centre $u \in K^n$ and radius r consists of all words $v \in K^n$ such that $d(u, v) \leq r$. The number of words in $B_q(u, r)$ will be denoted by $V(q, n, r)$.

PROPOSITION *Let K be a finite field of order q and let $u \in K^n$. Then the number of words in $B_q(u, r)$ is*

$$V(q,n,r) = \binom{n}{0} + \binom{n}{1}(q-1) + \cdots + \binom{n}{r}(q-1)^r = \sum_{k=0}^{r} \binom{n}{k}(q-1)^k.$$

Proof The ball $B_q(u, r)$ is the disjoint union of the sets of words v with $d(u, v) = k$ for $k = 0, \ldots, r$. To find a word at distance k we must first choose the k places at which v should differ from u. The number of ways of doing this is the binomial coefficient $\binom{n}{k}$. Then in each of these places we must choose a symbol different from the one u has in that location. There are $q - 1$ ways of doing this. All together that gives $(q - 1)^k$ possibilities once the places have been chosen. Thus there are $\binom{n}{k}(q - 1)^k$ words at distance k from u. □

15.2 The simple Gilbert bound

This bound follows from an easy counting argument using the number $V(q, n, r)$ defined above.

PROPOSITION *Let C be a linear code of block length n over the field K of order q with minimum distance d. If $|C| < q^n/V(q, n, d-1)$, then there exists a linear code C' strictly containing C in K^n such that the minimum distance of C' is still d.*

Proof The hypothesis implies that the balls of radius $d - 1$ centred on code words of C cannot cover the whole of K^n. So we can choose a word v such that its distance from all words of C is at least d. Construct C' as the set of all sums $u - av$, where $u \in C$ and $a \in K$. Then obviously C' is linear. To show that C still has minimum distance d, we must show that every non-zero code word w has weight $\mathrm{wt}(w) \geq d$. Let $w = u - av$ be a non-zero code word. Then $a = 0$ is equivalent to $w \in C$, and so in that case w has weight at least d by assumption. If on the other hand $a \neq 0$, let $b = a^{-1}$. Then

$$\mathrm{wt}(w) = \mathrm{wt}(bw) = \mathrm{wt}(bu - v) = d(bu, v).$$

Since C is linear $bu \in C$ and v was chosen to have distance at least d from all code words of C. Thus $d(bu, v) \geq d$. This establishes that all non-zero code words of C' have weight at least d as required. □

COROLLARY Simple Gilbert bound *For any n and $d \leq n$ there exists a code over the field of order q, which has block length n, minimum distance d, and rank at least $n - \log_q(V(q,n,d-1))$. In other words, the code has rate at least $1 - \log_q(V(q,n,d-1))/n$.* □

EXAMPLE For short block lengths many codes surpass the Gilbert bound. For example, the (8,7)-parity check code, which simply adds an overall parity check bit to each message word of length 7, has minimum distance $d = 2$, because it consists of all the words of even weight. The Gilbert bound for this case gives $|C| \geq 2^8/V(2,8,1) = 2^8/9$. That implies that there exists a linear code of block length 8 and minimum distance 2 with at least 32 code words. The parity check code has 128 code words.

15.3 The relative minimum distance

If we keep the minimum distance fixed as the block length increases, then the code becomes progressively weaker, since the ability to detect or correct a fixed number t of errors in a block of length n becomes less demanding and less useful as n increases. In comparing codes of differing block lengths we divide the rank and minimum distance by the block length.

DEFINITION For a code C of block length n, rank m, and minimum distance d, the *rate* or relative rank is $\mu = m/n$ and the *relative minimum distance* is $\delta = d/n$.

EXAMPLE If the relative minimum distance is kept fixed the Gilbert bound becomes harder to achieve as the block length increases.

Consider the double error-correcting code BCH(4,2), constructed in my book (Pretzel 1996). This has block length 15 and minimum distance 5. Its rank is 5 so its rate of 1/3 is much worse than that of the parity check code of the previous example. However, its relative minimum distance of 1/3 is somewhat better than the 1/4 of the parity check code. This code has 32 code words, but the Gilbert bound states that a linear code with these properties and at least 56 code words exists (see Exercise 1).

15.4 The entropy function

It is shown in my book (1996) that, at least as far as their designed parameters are concerned, long BCH codes fail to meet the Gilbert bound by an increasingly large margin. In order to make such asymptotic statements we shall also need an estimate for the second term of the Gilbert bound in which n no longer appears explicitly. That can be done by means of the q-ary entropy function.

DEFINITION For $0 \leq \delta \leq (q-1)/q$ we define the *q-ary entropy* function $H_q(\delta)$ by $H_q(0) = 0$ and

$$q^{H_q(\delta)} = (q-1)^\delta/\delta^\delta(1-\delta)^{1-\delta} \quad \text{or}$$
$$H_q(\delta) = \delta\log_q(q-1) - \delta\log_q(\delta) - (1-\delta)\log_q(1-\delta).$$

The following proposition describes the basic properties of the entropy function.

PROPOSITION (a) *For all* $0 < \delta < (q-1)/q$, *the function* $H_q(\delta)$ *satisfies* $H_q(\delta) \geq \delta q/(q-1)$ *with equality for* $\delta = 0$ *and* $(q-1)/q$.
(b) *The maximum value of* $H_q(\delta) - \delta$ *is attained for* $\delta = (q-1)/(2q-1)$. *It is* $\log_q(2q-1) - 1$.
Proof The derivative of $H_q(\delta)$ with respect to δ is $\log_q(q-1) - \log_q(\delta/(1-\delta))$, which is positive in the interval $0 < \delta < (q-1)/q$. The second derivative of $H_q(\delta)$ is $(\delta - 1)/(\delta \ln(q))$, which is negative. So the curve $y = H_q(x)$ is concave, constantly turning clockwise.

(a) For $\delta = (q-1)/q$, $H_q(\delta) = 1$. Hence the straight line joining the origin to $((q-1)/q, 1)$ must stay below the curve everywhere.

(b) The maximum of $H_q(\delta) - \delta$ is achieved when the derivative of $H_q(\delta)$ is $1 = \log_q(q)$ (it is a maximum because the second derivative is negative). That gives us an equation

$$\log_q(q-1) - \log_q(\delta/(1-\delta)) - \log_q(q) = 0.$$

Taking antilogs this becomes $(q-1)(1-\delta)/q\delta = 1$. The remainder of the proof is left as an exercise to the reader (Exercise 2). □

15.5 The relation between entropy and volume

With the help of this function we can replace the function $V(q, n, d-1)$ in the Gilbert bound.

LEMMA *Let* $0 \leq \delta \leq (q-1)/q$, *and for any integer* n *let* $r = r(n)$ *be the greatest integer such that* $r \leq \delta n$. *Then*
(a) $\log_q(V(q,n,r)) \leq nH_q(\delta)$ *and*
(b) *the limit of* $n^{-1}\log_q(V(q,n,r))$ *as* n *tends to* ∞ *is* $H_q(\delta)$.

Proof (a) Note that $0 \leq 1/q = 1 - (q-1)/q \leq 1 - \delta$. Hence for any $k \geq 0$,

$$\delta^k \leq (q-1)^k/q^k \leq (q-1)^k(1-\delta)^k.$$

Thus for $0 \leq i \leq \delta n$, taking $k = \delta n - i$ we get

$$\delta^{\delta n - i} \leq (q-1)^{\delta n - i}(1-\delta)^{\delta n - i}$$

and hence separating powers and multiplying by $(1-\delta)^n$,

$$\delta^i(1-\delta)^{n-i}/(q-1)^i \geq \delta^{\delta n}(1-\delta)^{n-\delta n}/(q-1)^{\delta n} = q^{-nH_q(\delta)}.$$

Now we write $1 = 1^n = (\delta + (1-\delta))^n$ and expand the right hand side using the binomial formula.

$$\begin{aligned}
1 = 1^n &= (\delta + (1-\delta))^n \\
&= \sum_{i=0}^{n} \binom{n}{i} \delta^i (1-\delta)^{n-i} \\
&= \sum_{i=0}^{r} \binom{n}{i} (q-1)^i \left(\frac{\delta}{q-1}\right)^i (1-\delta)^{n-i} \\
&\geq \sum_{i=0}^{r} \binom{n}{i} (q-1)^i \left(\frac{\delta}{q-1}\right)^{\delta n} (1-\delta)^{n-\delta n} \\
&= V(q,n,r) q^{-nH_q(\delta)}.
\end{aligned}$$

Taking logarithms to base q gives

$$0 \geq \log_q(V(q,n,r)) - nH_q(\delta),$$

proving (a).

(b) For the proof of (b) we need to apply Stirling's formula for $\ln(n!)$.

$$\ln(n!) - 1/12n \leq (n + 1/2)\ln(n) - n + K \leq \ln(n!),$$

where K is the constant $(\ln(2\pi)/2)$. If we convert to logarithms to the base q, the constant changes and we get

$$\log_q(n!) - \log_q(\mathrm{e})/12n \leq (n + 1/2)\log_q(n) - n\log_q(\mathrm{e}) + K' \leq \log_q(n!).$$

Certainly $V(q,n,r)$ is at least as large as any of the terms in the sum defining it. Thus

$$V(q,n,r) \geq \binom{n}{r}(q-1)^r.$$

We expand $\binom{n}{r} = n!/r!(n-r)!$, take logarithms to base q, and use Stirling's formula to estimate the factorials. Omitting subscripts q we get

$$\begin{aligned}
\log(V(q,n,r)) \geq (n+1/2)\log(n) &- (r+1/2)\log(r) \\
&- (n-r+1/2)\log(n-r) + r\log(q-1) \\
&- n\log(\mathrm{e}) + r\log(\mathrm{e}) + (n-r)\log(\mathrm{e}) - K' \\
&- \log(\mathrm{e})/12r - \log(\mathrm{e})/12(n-r).
\end{aligned}$$

When we divide by n and let n tend to ∞ we can ignore the terms that tend to 0 and the terms involving $\log(\mathrm{e})$ which cancel and get

$$\begin{aligned}
&\lim(n^{-1}\log(V(q,n,r))) \\
&\qquad \geq \lim\left(\log(n) - (r/n)\log(r) - ((n-r)/n)\log(n-r) + (r/n)\log(q-1)\right).
\end{aligned}$$

Now, if we choose r as the greatest integer satisfying $r \leq \delta n$, then as n tends to ∞ both $\delta n/r$ and $(1-\delta)n/(n-r)$ tend to 1. Hence

$$\begin{aligned}
&\lim(n^{-1}\log(V(q,n,r))) \\
&\quad \geq \lim(\log(n) - \delta\log(\delta n) - (1-\delta)\log((1-\delta)n) + \delta\log(q-1)) \\
&\quad = \lim(\log(n) - \delta\log(\delta) - \delta\log(n) \\
&\qquad\qquad - (1-\delta)\log(1-\delta) - (1-\delta)\log(n) + \delta\log(q-1)) \\
&\quad = \lim(-\delta\log(\delta) - (1-\delta)\log(1-\delta) + \delta\log(q-1)) \\
&\quad = H_q(\delta).
\end{aligned}$$

As $\log(V(q,n,r)) \leq nH_q(\delta)$ for all n, the limits must be equal. □

15.6 The asymptotic Gilbert bound

We can now prove Gilbert's theorem (Gilbert 1952).

THEOREM Asymptotic Gilbert bound *For all $\delta \leq (q-1)/q$, there exists a sequence C_n of linear block codes over F_q with block length of C_n equal to n, the relative minimum distance of C_n tending to δ, and rate tending to $1 - H_q(\delta)$.*

Proof Let r be the greatest integer satisfying $r \leq \delta n$. By the Gilbert bound there exists a linear code C_n over F_q of block length n, minimum distance $r+1$, and rate $m/n = 1 - n^{-1}\log(V(q,n,r))$. By Lemma 15.5 the limit of the rates of the codes C_n is $1 - Hq(\delta)$. To determine the limit of the relative minimum distances observe that $\delta n < r+1 \leq \delta n + 1$. So $\delta < (r+1)/n \leq \delta + 1/n$. Hence the limit of $(r+1)/n$ as n tends to ∞ is δ. □

15.7 Geometric Goppa codes and the Gilbert bound

Suppose we wish to construct geometric Goppa codes that meet the asymptotic Gilbert bound. The parameters of code $C_\Omega(B,D)$ for D of degree greater than $2\gamma - 2$ are

$$\begin{aligned}
\text{block length } n &= d(B), \\
\text{rank } m &\geq n - d(D) + \gamma - 1, \\
\text{min. distance } d &\geq d(D) - 2\gamma + 2.
\end{aligned}$$

We treat these estimates as if they were the true values. Then the following theorem gives the conditions under which a curve yields a good Goppa code.

THEOREM *Let C be a curve (or function field) defined over the finite field K of order q. Suppose that C has n rational places and genus γ. Then if*

$$\gamma \leq (\log_q(2q-1) - 1)n,$$

there is a geometric Goppa code defined on C whose design parameters meet the asymptotic Gilbert bound.

Proof To eliminate the term $d(D)$ from the estimates we write the relative minimum distance as $\delta = (d(D) - 2\gamma + 1)/n$. Then the rate satisfies $m/n = 1 - \delta - \gamma/n$. This will meet the Gilbert bound if

$$1 - H_q(\delta) \leq 1 - \delta - \gamma/n$$

or

$$H_q(\delta) - \delta \geq \gamma/n.$$

The maximum value for $H_q(\delta) - \delta$ was established in Proposition 15.4b. It is $\log_q(2q-1) - 1$. So for $\delta = (q-1)/(2q-1)$ the condition becomes $\gamma \leq n(\log_q(2q-1) - 1)$ as claimed. □

15.8 The theorem of Tsfasman, Vlăduţ, and Zink

In 1982 Tsfasman, Vlăduţ, and Zink, published a paper giving an 'explicit' description of a sequence of geometric Goppa codes over F_{p^2} whose rate and relative minimum distance tend to the asymptotic Gilbert bound. Their theorem is discussed in detail in the books by Tsfasman and Vlăduţ, (1991) and Moreno (1991). Their construction uses the so-called Shimura modular curves, and requires deep algebraic geometry. The curves used by Tsfasman *et al.* are examples of the following proposition, which I shall not prove.

PROPOSITION *For a finite field F_{p^2} with p prime ≥ 7, there exists a family of smooth curves such that the number n of rational points on the curves tends to infinity but γ/n tends to $1/(p-1)$ (where γ denotes the genus).* □

15.9 Towards a good family

It is straightforward to use the proposition above to derive the existence of a family of good geometric Goppa codes.

THEOREM *For q, the square of a prime $p \geq 7$, there exist one-point codes over F_q with block length n tending to ∞ such that for $\delta = (q-1)/(2q-1)$ the relative minimum distance of the codes tends to a limit $> \delta$ and their rate to a limit $> 1 - H_q(\delta)$.*

Proof Consider the sequence of curves of Tsfasman *et al.* and choose one of their rational points P. Let B be the sum of the other rational points and let $D = aP$. Let the genus of the curve be γ; then putting $d(B) = n$, we have $(\gamma - 1)/n \to 1/(p-1)$. Provided that $2\gamma - 2 < a$ the rank m and minimum distance d of the code satisfy

$$\begin{aligned} m &\geq n - a + \gamma - 1 \\ d &\geq a - 2\gamma + 2. \end{aligned}$$

Thus $m/n \geq 1 - a/n + (\gamma-1)/n$ and $d/n \geq a/n - 2(\gamma-1)/n$. Choose a so that $(a - 2(\gamma-1))/n \to \delta$. Then the limit of d/n is at least δ and the limit of m/n is at least

$$1 - \delta - \lim((\gamma-1)/n) = 1 - \delta - 1/(p-1).$$

Now for $p = 7$ we calculate $\log_{49}(97) \approx 1.1755 > 7/6$. Thus $1/(p-1) < \log_q(2q-1)$. As the left hand side of this inequality decreases with p and the right hand side increases, the inequality remains in force for larger p. So

$$1-\delta-\lim((\gamma-1)/n) > 1-\delta-\log_q(2q-1) = 1-H_q(\delta).$$

Thus the codes tend to the asymptotic Gilbert bound for this δ. □

15.10 The curves of Garcia and Stichtenoth

In their papers 1996, Garcia and Stichtenoth construct a sequence of curves by giving explicit equations that have good asymptotic properties. I shall present a special case of their construction here, in which the curves are defined in n dimensions over the field F_8.

EXAMPLE The equations for the curves are

$$x_2^7 = \sum_{j=0}^{6} x_1^j;$$

$$x_{i+1}^7 = \sum_{j=1}^{7} x_i^j x_1^{7-j} \qquad \text{for } i = 2, \ldots, n-1.$$

The projective coordinate transformation $y_1 = 1/x_1$, $y_i = x_i/x_1$ for $i > 1$ converts these equations to

$$y_{i+1}^7 = \sum_{j=1}^{7} y_i^j \qquad \text{for } i = 1, \ldots, n-1.$$

Although these equations are simpler they have the disadvantage that most of the rational points of the curves lie on the y-horizon, whereas there is a unique point on the x-horizon. So we describe the rational points in the x-system. There are

7^{n-1} points of the form $x_1 = 0$, $x_i \neq 0$ arbitrary for $i > 1$.

$6n$ points of the form $x_i \neq 0, 1$, $x_j = 1$ for $j < i$, $x_j = 0$ for $j > i$.

One point at infinity P. This is the point with y-coordinates $(0, \ldots, 0)$.

Thus the total number of rational points is $7^{n-1} + 6n + 1$.

The partial derivatives of the $n-1$ polynomials with respect to the n variable $y_1, \ldots, y_n$ at P form an $(n-1) \times n$ matrix. For the point $(0, \ldots, 0)$ this matrix has full rank and so that point is simple (see Exercise 4). A very rough estimate of the genus of these curves can be obtained by searching for polynomials in $L(aP)$ for large a.

From the equations of the curve in the y-system one can derive the fact that $(y_{i+1}^7/y_i)(P) = 1$ and hence

$$\nu_P(y_i) = 7^{n-i}b,$$

where $b = \nu_P(y_n)$. The fact that some function f must have $\nu_P(f) = 1$ implies that $b = 1$. Now it follows that

$$\nu_P(x_1) = -7^{n-1}$$
$$\nu_P(x_i) = 7^{n-i} - 7^{n-1} \quad \text{for } i > 1.$$

We can calculate the possible values $\nu_P(f)$ for monomials $f = \prod x_i^{r_i}$. That gives an estimate for $l(aP)$ as the number of values $m \leq a$ for which an f with $\nu_P(f) = m$ exists. Call the values m for which no such polynomial exists *gaps.* They can be found by a straightforward search. Thus for $n = 2$ there are 15 gaps of which the largest is 29; for $n = 3$ there are 246 gaps, the largest being 491; and for $n = 4$ there are 2745 gaps, the largest being 5489. That gives estimates for the genus $\gamma_2 \leq 15$, $\gamma_3 \leq 246$, and $\gamma_4 \leq 2745$.

The curves are not smooth, so the horizon theorem cannot be applied and thus this type of calculation cannot easily be extended to find a complete basis of $L(aP)$.

Using stronger techniques, Garcia and Stichtenoth give a sharper estimate of the genus: $\gamma_n \leq 3\,(7^{n-1} + 6n + 1)$. That estimate allows the construction of codes that come close to the asymptotic Gilbert bound, though they do not quite meet it.

PROPOSITION *For the curves of Garcia and Stichtenoth there exists a sequence of Goppa codes C_k such that the block lengths of C_k tend to infinity, the relative minimum distance tends to $\delta = 7/15$, while the rate tends to a value above $1 - 1.1H_8(\delta)$.*

Proof Consider the sequence of curves of Garcia and Stichtenoth. Let B be the sum of the finite rational points and let $D = aP$, where P is the point at infinity. Let the genus of the curve be γ, and put $d(B) = n$. Provided that $2\gamma - 2 < a$ the rank m and minimum distance d of the code satisfy

$$m \geq n - a + \gamma - 1$$
$$d \geq a - 2\gamma + 2.$$

Thus $m/n \geq 1 - a/n + (\gamma - 1)/n$ and $d/n \geq a/n - 2(\gamma - 1)/n$. Choose a so that $(a - 2(\gamma - 1))/n \to \delta$. Then the limit of d/n is at least δ and the limit of m/n is at least

$$1 - \delta - \lim((\gamma - 1)/n) \geq 1 - \delta - 1/3.$$

The value $H_8(\delta) - \delta$ is $\log_8(15) - 1 \approx 0.3203$. Hence $\delta + 1/3 \approx 1.04H_8(15)$. It follows that the limit of m/n is greater than $1 - 1.1H_8(15)$. □

Although these curves are much more 'concrete' than those of Tsfasman *et al.*, their precise genus has not yet been calculated, and as a result, the codes promised by the proposition can still not be constructed explicitly.

15.11 Exercises

1. Use the Gilbert bound to show that there exists a linear binary code of block length 15, minimum distance 5, and rank 6.

2. Conclude the proof of Proposition 15.4 by showing that $(q-1)(1-\delta)/q\delta = 1$ implies that $\delta = (q-1)/(2q-1)$ and evaluating $H_q(\delta) - \delta$ for this value.
3. For the curves of Garcia and Stichtenoth in Paragraph 15.10, find the defining equations in the y-system, where $y_1 = 1/x_1$ and $y_i = x_i/x_1$ for $i > 1$ and verify that the only point P with $y_1 = 0$ is $y_i = 0$ for all i.
4. It is possible to show for a curve in n dimensions defined by $n-1$ polynomials $f_j(x)$ that a point is non-singular if the rank of the matrix of the partial derivatives $(\partial f_j/\partial x_i)$ is $n-1$. Verify that the point P is non-singular.
5. Verify the statements about values $y_{i+1}^7/y_i(P)$ and the values $\nu_P(y_i)$ in Paragraph 15.10. Deduce the claimed values for $\nu_P(x_i)$ and show that these functions have no further poles. Finally confirm the calculations for the gap numbers of values for monomials x^r.

15.12 Research problems

I conclude the book with some research problems.

1. Develop a simpler theory of Goppa codes
A start in this direction has been made (Feng *et al.* 1994), but it deals only with particular codes. Their work suggests that it might be possible to prove the Riemann–Roch theorem by proving that a decoding algorithm works rather than vice versa. That might make it possible to avoid the introduction of repartitions and differentials altogether. However, it is more likely that this approach will remain restricted to particular classes of codes, for which a direct construction is known.
2. Develop efficient, practical error-correcting algorithms for Goppa codes
There has also already been some progress on this problem by Sakata and the school of Justesen (Sakata *et al.* 1995), using Sakata's multi-dimensional extension of the Berlekamp–Massey algorithm. It is also possible that an efficient algorithm might emerge from the work of Porter *et al.* (1992).
3. Determine the exact genus of the curves of Stichtenoth and Garcia.
4. Find other explicit families of curves yielding good Goppa codes and determine their genera.
5. Develop a theory of *partial* Goppa codes
This is an attempt to evade the exact calculation of the genus. You are given a basis of a subspace of $S \subset L(D)$ and construct the partial code $C_\Omega(B, D, S)$ consisting of all words c such that $\varphi \cdot c = 0$ for all $\varphi \in S$. This code contains the code $C_\Omega(B, D)$ as a subcode. Find conditions on the basis the allow estimates for the rank and minimum distance of $C_\Omega(B, D, S)$. Extend the decoding algorithms to these codes.

REFERENCES

1. Artin, E (1967). *Algebraic numbers and algebraic functions.* Gordon and Breach, New York.
2. Birkhoff, G. and Maclane, S. (1977). *A survey of modern algebra* (4th edn). Macmillan, New York.
3. Chevalley, C. (1951). *Introduction to the theory of algebraic functions of one variable.* American Mathematical Society, Providence, RI.
4. Cohn, P. M. (1977). *Algebra* Vol. 2. John Wiley & Sons, Chichester.
5. Cohn, P. M. (1982). *Algebra* Vol. 1 (2nd edn). John Wiley & Sons, Chichester.
6. Cox, D., Little, J., and O'Shea, D. (1992). *Ideals, varietes, and algorithms. An introduction to computational algebraic geometry and commutative algebra.* Springer Verlag, New York.
7. Deuring, M. (1973). *Lectures on the theory of algebraic functions of one variable* Lecture Notes in Mathematics 314. Springer Verlag, Berlin.
8. Duursma, I. M. (1993). Majority coset decoding. *IEEE Trans. Inf. Theor.* **39**, 1067–1070.
9. Feng, G.-L. and Rao, T. R. N. (1993). Decoding algebraic-geometric codes up to the designed minimum distance. *IEEE Trans. Inf. Theor.* **39**, 37–45.
10. Feng, G.-L., Wei, V. K., and Rao, T. R. N. (1994). Simplified understanding and efficient decoding of a class of algebraic-geometric codes. *IEEE Trans. Inf. Theor.* **40**, 981–1002.
11. Fulton, W. (1989). *Algebraic curves, an introduction to algebraic geometry* Reprint of 1969 edition. Addison-Wesley, Redwood City, CA.
12. Garcia, A. and Stichtenoth, H. (1995a). A tower of Artin-Schreier extensions of function fields attaining the Drinfeld-Vlăduţ bound. *Invent. Math.* **121**, 211–222.
13. Garcia, A. and Stichtenoth, H. (1995b). Algebraic function fields over finite fields with many rational places. *IEEE Trans. Inf. Theor.* **41**, 1548–1563.
14. Garcia, A. and Stichtenoth, H. (1996). Asymptotically good towers of function fields over finite fields. *C. R. Acad. Sci. Paris* **322 I**, 1067–1070.
15. Gilbert, E. N. (1952). A comparison of signalling alphabets. *Bell Syst. Tech. J.* **31**, 504–522.
16. Goppa, V. D. (1981a). Codes on algebraic curves. *Dokl. Akad. Nauk. SSSR* (Russian) **259**, 1289–90.
17. Goppa, V. D. (1981b). Codes on algebraic curves. *Soviet Math. Dokl.* (translation) **9**, 207–14.
18. Justesen, J., Larsen, K. J., Havemose, H. E., Jensen, H. E., and Høholdt, T. (1989). Construction and decoding of a class of algebraic geometry codes. *IEEE Trans. Inf. Theor.* **35**, 811–821.

19. Justesen, J., Larsen, K. J., Jensen, H. E., and Høholdt, T. (1992). Fast decoding of codes from algebraic plane curvese. *IEEE Trans. Inf. Theor.* **38**, 111–119.
20. Lint, J. H. van and Geer, G. van der (1988). *Introduction to coding theory and algebraic geometry.* Birkhäuser, Basel.
21. Moreno, C. (1991). *Algebraic curves over finite fields.* Cambridge Univerity Press, Cambridge.
22. Porter, S. C., Shen, B.-Z., and Pellikaan, R. (1992). Decoding geometric Goppa codes using an extra place. *IEEE Trans. Inf. Theor.* **38**, 1663–1676.
23. Pretzel, O. (1992). *Error-correcting codes and finite fields.* Oxford University Press, Oxford.
24. Pretzel, O. (1996). *Error-correcting codes and finite fields* (Student Edition). Oxford University Press, Oxford.
25. Roquette, P. (1958). Über den Riemann-Rochschen Satz in Funktionenkörpern vom Transzendenzgrad 1. *Math. Nachr.* **19**, 375–404.
26. Sakata, S., Jensen, H. E., and Høholdt, T. (1995). Generalized Berlekamp–Massey decoding of algebraic-geometric codes up to half the Feng–Rao bound. *IEEE Trans. Inf. Theor.* **41**, 1762–1768.
27. Skorobogatov, A. N., and Vlăduţ, S. G. (1990). On the decoding of algebraic-geometric codes. *IEEE Trans. Inf. Theor.* **36**, 1051–1960.
28. Stichtenoth, H. (1978). *Algebraische Funktionenkörper einer Variablen.* Vorlesungen aus dem Fachbereich Mathematik. Universität Essen.
29. Stichtenoth, H. (1991). *Algebraic function fields and codes.* Springer Verlag, Berlin.
30. Tate, J. (1952). Genus change in inseparable extensions of function fields. *Proc. Am. Math. Soc.* **3**, 400–406.
31. Tsfasman, M. A. and Vlăduţ, S. G. (1991). *Algebraic-geometric codes.* Kluwer, Amsterdam.
32. Tsfasman, M. A., Vlăduţ, S. G., and Zink, T. (1982). Modular curves, Shimura curves and Goppa codes better than the Varshamov-Gilbert bound. *Math. Nachr.* **109**, 21–28.
33. Varshamov, R. R. (1957). Estimate of the number of signals in error-correcting codes. *Dokl. Akad. Nauk. SSSR* **117**, 739-741.

INDEX